Frederick Colyer

Treatise on Water Supply, Drainage, and Sanitary Appliances of Residences

Including Lifting Machinery, Lighting and Cooking Apparatus, etc.

Frederick Colyer

Treatise on Water Supply, Drainage, and Sanitary Appliances of Residences
Including Lifting Machinery, Lighting and Cooking Apparatus, etc.

ISBN/EAN: 9783743401495

Manufactured in Europe, USA, Canada, Australia, Japa

Cover: Foto ©berggeist007 / pixelio.de

Manufactured and distributed by brebook publishing software
(www.brebook.com)

Frederick Colyer

Treatise on Water Supply, Drainage, and Sanitary Appliances of

Residences

TREATISE

ON

WATER SUPPLY, DRAINAGE,

AND

SANITARY APPLIANCES OF RESIDENCES;

INCLUDING LIFTING MACHINERY, LIGHTING AND
COOKING APPARATUS, &c.

By FREDERICK COLYER,

M. INST. C.E. ;

AUTHOR OF 'BREWERIES AND MALTINGS: THEIR ARRANGEMENT, ETC.';
'HYDRAULIC AND STEAM LIFTING MACHINERY'; 'MODERN STEAM
ENGINES AND BOILERS'; 'PUMPS AND PUMPING MACHINERY';
'CONSTRUCTION OF GAS WORKS'; 'MANAGEMENT OF
ENGINES AND BOILERS'; 'PUBLIC INSTITUTIONS,
THEIR ENGINEERING AND OTHER APPLIANCES, ETC.'

E. & F. N. SPON, 125, STRAND, LONDON.

NEW YORK: 12, CORTLANDT STREET.

1889.

PREFACE.

IT was suggested to me by a professional friend that a book describing "Sanitary Work, Cold and Hot Water Supply, Warming Apparatus, &c., for Residences," was much wanted, and if written in a concise form might be acceptable to architects and surveyors who may not have had much experience in matters of this kind. The greater part of the work described is the result of experience in my own practice in carrying out both architectural and engineering works. Should this small treatise supply a want and be found useful, it will give me much gratification to have been of any assistance to my brethren. I shall at all times be pleased to be of any service to any professional man, if he will do me the favour to communicate on any matter I have not succeeded in making clear, or that may be wanting in sufficient detail.

FREDERICK COLYER, M.I.C.E.,
Civil Engineer and Architect.

18, GREAT GEORGE STREET,
WESTMINSTER, S.W.

CONTENTS.

———◆———

CHAPTER I.

INTRODUCTION.

CHAPTER II.

SANITARY ARRANGEMENTS.

CONTENTS.

CHAPTER III.

DRAINAGE.

CHAPTER IV.

WATER SUPPLY FOR HOUSEHOLD PURPOSES.

CHAPTER V.

WARMING APPARATUS.

CHAPTER VI.

GAS AND ELECTRIC LIGHTING, BELLS, ETC.

Gas pipes and fittings, proper system to fix same—Meter, best position—Size of pipes—Outside supply mains— Necessity for first-class work—Ventilation to get rid of products of gas combustion—Inlets for fresh air—Boyle's ventilators—Gas stoves—Gas fires. *Electric lighting*— Large instalments for mansions — Leading details of machinery and plant—Boilers, engines, dynamos; data for same—Incandescent lights—"Arc" lights—Wires and safety fuses—Power Co.'s supply—Power and price of lamps— Number wanted to light. *Electric bells*—Batteries, wires, &c.—Indicators and signals. *Telephones.*

CHAPTER VII.

LIFTING MACHINERY.

Food and coal lifts—Hand and hydraulic power—Passenger lifts—Details of construction—Best kind to use — Proportions — Way of working — Lifts for invalids — Water supply to lifts—Various kinds of lifts—Reference for further details, &c.   ..   ..   ..

CHAPTER VIII.

GAS AND STEAM COOKING APPARATUS.

Gas-ovens — Frying-plates — Broiling-stoves — Steam-ovens —Jacketed copper — Vegetable kitchen, &c. — Bakery— Machines for making bread—Ovens: hot water, gas, and fire-heated—Construction of bakeries—Material for floors —Ventilation   ..   ..

CONTENTS.

CHAPTER IX.

LAUNDRIES.

CHAPTER X.

GREENHOUSES, HOTHOUSES, ETC.

TREATISE

ON

WATER SUPPLY, DRAINAGE,

AND

SANITARY APPLIANCES OF RESIDENCES.

———◦◦———

CHAPTER I.

INTRODUCTION.

In modern residences of the present day architects
have to deal with a large number of special appliances and apparatus. It has been suggested to the
author that a few hints as to the best way of successfully carrying out these might be acceptable to
many. Being engaged in the joint practice of
architecture and engineering, he is well acquainted
with the requirements for residences and other
buildings.

The cold and hot water supply, heating apparatus, and sanitary works, require, in most cases,
special provision made in the building, a proper
consideration of which so much adds to the comfort
and convenience of the occupiers.

It is to be feared in many instances the works
above named do not receive the attention they
deserve from many architects, and are too often left

to tradesmen and workmen, who are ignorant as to the proper and efficient way of carrying out such works. The author is quite aware that the detail in most cases is very troublesome, and that the architect is, as a rule, very inadequately remunerated for this part of the work; he is, however, certain these important matters can only be successfully carried out by detailed description in specifications and by close *personal* supervision afterwards. In his own practice the author seldom finds, when making surveys of residences for clients, that the important things above named have received the attention they deserve. This remark especially applies to the drainage; in too many cases the plumbing and other work is of the worst description, showing not only ignorance, but in too many instances the manner in which the work is done is disgraceful. It is well known that details of this kind are very distasteful to many designers; the importance, however, of making residences healthy and fit to live in should outweigh every other consideration, and induce the architect to give it the attention it deserves.

It would be impossible to indicate in detail what is essential to suit every case; the leading points of the specialities before named will therefore be principally touched upon, and hints given as to the best course to pursue.

As the sanitary arrangements of a house are one of the most important things, they will be first treated; drainage will follow; then cold and hot water supply, warming apparatus, &c.

OLD RESIDENCES.

With regard to the condition of old houses, or those built in the last few years, the sanitary arrangements for water supply, and other details upon which the comfort of the inmates depends, should be made the subject of the most careful survey. This applies more particularly to country houses, especially those which have been built a quarter of a century and upwards; the author has seldom found that houses which he has surveyed were in a proper sanitary condition and perfectly healthy. In making a survey of such places, attention should be given to the following matters.

The water supply should be one of the chief considerations: first examine the source of supply, and be sure that perfectly pure water is obtained; if the source is from wells, examine their position, make sure that the surface water is excluded, and that no drainage, either surface water or sewage, can get in to contaminate the water. It may surprise some to hear that in old mansions in the country, the cesspool and the well are often found in dangerous proximity; thus in one instance the author found upon inquiry that the exact position of the well was not known—no one had seen it for seven years! It was in a back-yard, covered over, and suspiciously near to some very dirty cow-sheds and piggeries; part of the sewage taken out from the house in old barrel drains was known to run very near it; it was also believed an old cesspool was very close to it. The only source of water supply was from this well; it was pumped up into a cistern which was fixed over the water-closet, and from this all the water was taken, both for water-closets and for drinking and cooking purposes. The house drainage was of the worst description. If such a condition of things existed in a fair-sized mansion, it is probable worse things would be found in smaller houses.

The next important matter to examine is the whole system of drainage. Have all necessary parts of the drains uncovered, also make sure there are no old disused barrel or pipe drains under the floors; see how the water-closets, baths, lavatories, and sinks are connected, where they are ventilated, and in what condition the fittings are, as to close joints, &c. Particular attention should be given to all waste water pipes, especially those that run from baths and lavatories or housemaids' sinks, to see that they are absolutely disconnected from all soil pipes or water-closet drainage. The greatest care-lessness and ignorance is often displayed, especially by country workmen, in the connections made; it has too often, in the author's experience, been the cause of serious illness and even death. The kitchen drainage, and water supply to same for cooking and cleaning purposes, should have close attention; in too many instances the sinks are of the most miserable description, and the outlet for waste water is made direct into the drains, and in many cases close to the outlet of the soil pipe. Bell-traps in sinks should be absolutely abolished; when left open, as they usually are for a good part of the day by careless servants, the house is in direct communication with the foul air from the drains. The heat from the kitchen or scullery fire draws up the sewer gas, which not only contaminates the food while cooking, but poisons the air of the house.

It would take too much time to read all the experiences the author has had in making different surveys ; he fears such details would be both tedious and unpleasant. He, however, ventures to advise, not only professional men, but tenants of houses, to have all their domestic and sanitary arrangements carefully surveyed by some one skilled in these matters, and not to trust to tradesmen, who are too often ignorant in such essential things. The author ventures to submit the following general rules as a safe guide to ensure healthy houses :—

1. The water supply must be divided into two systems : one for drinking, cooking, and washing ; and one for the water-closets. No connection of any kind to be made between the two services.

2. The water-closets and sink from the kitchen to discharge into one system of drainage, the waste water from baths and lavatories into another, and the rain water from the roofs into a third.

3. All *direct* connections between the house and the drains should be *absolutely* cut off.

4. All drain and soil pipes should be well venti-lated, and fresh air inlets provided. The pipes and syphon traps must be so arranged that they can be easily cleansed and any stoppage removed.

5. All overflow pipes from water tanks should deliver into the open air outside the house. *No*

connection should ever be made between the overflow pipes from tanks and any water closet.

Details are given, under their various heads, of the best way to carry out all these matters, to which the reader is referred. The author may, perhaps, be permitted to say, the manner proposed for doing the work is not the result of theory, but what he has proved from experience in his practice to be the proper way of carrying out the works described.

The author lays special stress upon water supply and sanitary work, because it is such a highly important matter, and one that he fears is too often neglected by many architects.

Warming, lighting, cooking appliances, and laundry work, together with sundry other essential things, have been treated in as much detail as the limits of this work would permit. Most essential fittings and appliances of any moment have been discussed, it may be more minutely than some require; indulgence must be given, when this is the case, by those who are well experienced in any particular specialities. It is hoped the information given will be useful to many, and be a guide to young professional men when they are designing new works, or have to make surveys of old or existing buildings.

The author is now publishing another work,* in which are described the sanitary and engineering works required in large public institutions. The subjects treated are necessarily more extended than in the present treatise.

* 'Engineering and other Appliances of Public Institutions,' by F. Colyer. E. & F. N. Spon, London.

CHAPTER II.

SANITARY ARRANGEMENTS.

UNDER this head will be treated drainage, water-closets, baths, lavatories and sinks, &c. It has been thought the matter will be made more clear if the apparatus is described first, and the drain pipes and sewers after, as it will better indicate the purpose the latter have to serve. It is somewhat difficult to divide some parts to save repetition in other sections; this, however, will be avoided as much as possible.

WATER CLOSETS.—These should always be placed next an outer wall where direct ventilation can be obtained, permanent communication with the outer air by means of ventilators is *absolutely* necessary. The size of the room for closets in a good residence should not be less than 3 feet 6 inches wide, 5 feet 6 inches long, 9 feet high, as a minimum.

The old style of closet, made with a pan, hopper or container, and D trap, should never be used; they are both dirty in action and are most insanitary, it is impossible to keep them properly cleansed; they are usually receptacles for all sorts of abominations and filth, as well as for foul air and sewer gas—their use cannot be too strongly

condemned. When the closet is removed for repairs or renewal, it is often found that the D trap
is eaten through at the side, and that the joints are
leaky, this was actually the case in works lately
carried out by the author where the client considered it perfect before it was taken out.

Closet Apparatus.—This should be *of the valve
kind*, and obtained from the most approved
makers ; the author has found those made by
G. Jennings, Dent & Hellyer, and Tylor & Sons,
the most efficient. None but the very best manufacture should be adopted, the first expense will be
found the cheapest in the end. The pan and trap
is best made in glazed earthenware, the trap or
syphon should be placed above the floor line. The
floor on which the pan stands should have a lead
safe, 6 lb. per super foot ; it should be made the
whole width of the closet enclosed by the woodwork of seat, the lead should be turned up to
form a skirting all round, not less than 3 inches
deep, and be secured to the woodwork by copper
nails. At one corner of the safe, a 1½ inch
diameter lead drain pipe, 7 lb. per foot, should be
provided to carry off any waste water ; the end of
the pipe should be taken through an outer wall,
and be provided with a copper flap set at an angle.
The soil pipe syphon trap should be solid drawn,
3½ to 4 inch diameter, and of 7 lb. lead. Wherever
possible, the down pipe should be placed outside
the house, and be supported at each 5 to 6 feet by
double lead tacks, 9 lb. per super foot ; these should

be fastened to wood blocks spiked to the wall. The soil pipe should discharge into a special ventilated trap at the bottom, to be hereafter described.

The soil pipe should be continued from the highest point of the discharge into it, above the top of the roof, to act as a ventilator. A Banner's patent cowl should be fixed at the top of the pipe. A 2-inch lead pipe should also be connected at the top of the syphon under the pan, to carry away any foul air ; this may be connected into the large vertical ventilating pipe ; an air inlet pipe is also provided, described under "Drainage." The soil pipe must be felted, and enclosed in wood casing, to protect it from the weather, as in some instances in very cold weather it is liable to freeze.

The junction between the flange of the syphon of the pan and the soil pipe must be made in the following way : the soil pipe must be brought through the lead safing at the floor, turned over as a flange, and then the flange of the syphon screwed through it to the floor ; by this means an absolute joint is made, which can be readily broken, when required to move the pan. It will be observed in this system the old abominable putty joint is not used ; it is to be hoped it will soon cease to exist.

Wood Fittings.—The riser, seat, and flap, &c., should be made of 1-inch mahogany ; the seat, as well as the flap, should be hung on brass butts, and the riser made to open in the same way ; this gives instant access to the pan and valves, and also enables the domestics to keep the under part of the closet clean. This is a very essential thing, and a matter too often neglected, the faint smell in some closets arises from this, and not from sewer air.

The Water Supply Pipe should be taken from the closet cistern ; it should be $1\frac{1}{4}$ to $1\frac{1}{2}$ inch diameter, of lead 6 lb. per foot; this ensures that an ample quantity of water is *quickly* discharged. Special attention is called to this, as it often happens that plenty of *pressure* is given, but not the *quantity* necessary to clean the pan and trap. In the event of water closets being required on other floors, it is the best plan, where possible, to place them in the same line, that is, one over the other, and to use the same soil pipe ; when more than one closet discharges into the same soil pipe, the main pipe down should not be less than 4 inches diameter.

The walls of the closet should be lined with glazed tiles, and, if possible, the floor may be covered with tiles ; the ceiling may be painted, to allow it to be washed occasionally.

When the position of the closets can be chosen, they should be placed so that they may be partially disconnected from the house ; this may be done by forming a lobby or corridor in front of each, provided with a window to open, or a ventilator in the outside wall. Should any foul air escape from the closet, it will be carried out of this window or ventilator, and not be taken into the house. In all houses of fair size it is advisable to provide separate closets for gentlemen ; those for ladies should be placed as a rule on the same level as the bed-room floor, and in cases where a heating apparatus is provided, the closet may be heated by a small coil of pipes. In houses where there are children and invalids this is a matter of some importance, and deserves attention. The convenient and private position of all closets is a matter demanding more attention than it usually receives.

Water-closets for Servants. — These should be placed outside the house; the size of the rooms should not be less than 3 feet wide by 4 feet 9 inches long by 7 feet 6 inches to 8 feet high. The floor should be covered with granolithic paving, the walls rendered in cement, or they may be covered with glazed tiles set in cement.

The closet apparatus should be of the Pedestal or Vase kind; the seat should be hinged, so as to use the pan as a slop hopper; the lower part of the seat should be left open. The water supply should be by a flushing cistern, holding at least three gallons, the supply pipe 1½ inch diameter. A ventilating pipe should be connected with the top of the syphon, and carried up the side of house, quite clear of all windows. The soil should be discharged into an iron trapped gulley, as described at p. 22. An air-inlet pipe and talc valve should be provided and fixed in the way described under the head of "Drainage." A ventilator that cannot be closed should be formed in the wall; this ensures a constant stream of fresh air. The water must be taken from the separate water closet service supplying

the inside closets.   The room should be well
lighted ; it is too often the case that these necessary conveniences are dark, with the consequent
result that they are generally dirty also.

The Long Hopper Closets for servants should
never be used ; they cannot be kept clean, and as
a rule they fall into a filthy condition.   The water
supply to these closets is usually very inefficient,
the syphon traps at the outlet from the pans are
often very faulty.

Earth Closets.—Where a good supply of water
can be had, and a cesspool conveniently near, this
kind of closet cannot be recommended.   Unless
they receive the most careful attention, they
become very foul and offensive.   Apparatus that
require any special attention give trouble, and so
too often get neglected, and are thus allowed to
get into a foul and filthy state.

Where there are male servants in the house, the
closets for their use should be placed in the garden
or yard, away from the house, in a private place.
The same kind of apparatus and method of fitting
up should be adopted as the last described.   Close
to the water closets for men, a urinal, made of
earthenware in a slate enclosure, should be provided,
and water laid on to keep basin clean.   The back
and sides may be of 1 inch slate slab ; it should be
fitted with a water spreader, and provided with a
trapped gully at the bottom to take away waste
water, &c.

URINALS

Should be provided in gentlemen's closets; the basins should be earthenware, and fitted with water supply. The discharge may be made into the soil pipe, but not into the syphon of the closet; they should be trapped at the outlet pipe, and also at the basin. G. Jennings and Dent & Hellyer make very good apparatus of this kind.

LAVATORIES.

These should be placed either in the bath rooms or in some equally convenient position; it is not desirable to place them in water-closet rooms, when possible to avoid it.

The Basins should be china, and either of the " tip " kind or made with discharge plugs, the latter as a rule, are most suitable for a private house. The waste water pipes should be lead, 7 to 8 lb. per foot, with a patent solid drawn lead trap, provided with cleaning cap, and placed immediately under the basin; the pipes should be $1\frac{1}{2}$ inch diameter, and taken through the wall of the house and made to discharge *over* a R.W.P. head. *They should never, on any account, be connected into any of the soil pipes,* this rule cannot be too rigidly enforced. The hot and cold water pipes and cocks should be $\frac{3}{4}$ inch diameter, the latter of the solid bottom packed-gland type.

The wood of the lavatory basin enclosure should be made to open in the front, to examine the trap and clean same when necessary. The water should be taken from the drinking water supply. The floor within the enclosure should have a lead safe 6 lb. per superficial foot, with skirting 3 inches deep nailed to woodwork all round, and $1\frac{1}{2}$ inch waste pipe in one corner, carried through the outer wall, and discharged over the R.W.P. head above named ; the end of the pipe should be fitted with a flap made of copper. When several lavatories are connected into one down or discharge pipe, it should not be less than $2\frac{1}{2}$ inch diameter. When these discharge pipes are more than 10 feet long, they should be carried up above the roof, and fitted with a cowl as described for soil pipes; this is for the purpose of ventilation. Dirty soapy water, if discharged into long lengths of pipes, often makes them very foul if not freely ventilated in the same way as described for soil pipes.

BATHS.

The rooms should not be less than 7 feet by 7 feet 6 inches by 9 feet high as a minimum ; a good size room is 8 feet 6 inches by 11 feet by 9 feet 6 inches high. A ventilator should be fixed in the outer wall, if possible away from the bath, so as to prevent any current of cold air coming on the bather.

Baths may either be made of copper, wrought iron, zinc, or fire-clay, the size 5 feet 6 inches to 6 feet long by 2 feet wide by 1 foot 10 inches deep. When of copper they should be tinned inside and made of metal 2 lb. per superficial foot. When of wrought iron, they should be No. 16 gauge (about $\frac{1}{16}$ inch thick) and enamelled inside.   Cast-iron baths are not to be recommended, as they are liable to crack, and the enamel wears off them.

The inlet pipes for cold water should be 1 inch diameter of lead, the hot water pipe 1 inch diameter of wrought iron, the waste pipes lead, 2 inches diameter and 9 lb. per foot ; the waste pipes should be taken through the wall and discharged over a R.W.P. head ; at the outlet a patent solid drawn lead trap, fitted with cleaning screw, should be provided and fixed immediately under the baths. The inlet and waste water pipes should *always* be kept separate.   The inlet water valves may be coupled to form a mixing chamber, and should be placed at the top of the bath.   A lead safe of 6 lb. per superficial foot should be provided under the bath over the whole area, turned up 3 inches deep and nailed to the skirting.   A lead pipe $1\frac{1}{2}$ inch diameter should be provided to take away the waste water ; it should pass through the wall and discharge over the R.W.P. head above named.

The weight of a fire-clay bath is about 6 cwt., these are very good baths to use if put in during the construction of the house.   As they are very heavy, they require special provision in the bearers to carry them.   They are somewhat more expensive than wrought iron, but cost rather less than tinned copper baths.

The Wood Enclosure should be 1 inch thick, panelled and made to open at the side for examination and cleaning.   The top frame of the enclosure should be mahogany $1\frac{1}{8}$ inch thick, properly framed ; a hinged cover is not necessary.

The hot water cistern may be placed on the floor of the bath room and be contained in a hot cupboard. The enclosure may be panelled and fitted with two doors, one at the level of the floor and one at a higher level; open hard wood grids should be provided and a hanging rail for the wet towels. The cistern should not be less than $\frac{3}{16}$ inch thick, made of galvanised iron ; the contents should be about 50 gallons. The hot water connections will be described under the head of "Hot Water Apparatus." In cases where the hot cistern is placed in the kitchen, see "Hot Water Apparatus," a coil of hot water pipes may be taken round the cupboard, or a small heater may be provided for drying towels and linen.

Fixing Baths.—The bath should rest upon bearers 3 inches to 4 inches above the floor. The height from the floor to the top of bath may be 2 feet 4 inches to 2 feet 9 inches as a maximum.

The pipes should be placed so that they can be easily got at for repairs when necessary, and should be provided with cleaning caps at the bends and junctions. The pipes should never be bedded in the wall or plaster, nor should they be placed under the boards ; no sharp bends or square tee-pieces should be used, or any junction made with other pipes. Gun metal unions should be used, to enable the connections to be easily taken apart. The weight of the lead pipes should not be less than stated ; it is false economy to put in light pipes or any inferior fittings, as they invariably give trouble after working a short time.

A *Shower Bath* may be arranged at one end of the bath, by carrying up the woodwork and enclosing it at the sides and end and hanging curtains in the front. The metal cistern with perforated bottom should be of copper and carried on the framing at top ; the height will be governed by the room.

A *Lavatory* may be placed in the bath room, fitted with hot and cold water service, and in the same way as described at p. 13. The waste water pipe should be taken through the wall and discharged over the same R.W.P. head that takes the bath water. A patent lead trap, fitted with cleaning screw, should be provided, and fixed immediately under the basin. A lead safing, 6 lb. per foot, and a skirting should be formed, and nailed to woodwork as described for the bath ; a waste water pipe must also be provided to take away any overflow or leakage. The lavatory basin in this case may be of extra size and fitted with cold and hot water jets for rinsing the hair, &c. A glass should be fixed at the back. The basin may be either made with a plug and side overflow pipe, or of the "tip" kind. When of the latter class, great care must be used in the trapping. Jennings' patent tip basins are the best of this class, they are carefully fitted up and are easily taken out for cleaning the under side and to get at the top of the trap. The room may be paved with tiles, and the walls may also be covered

C

in the same way. The lower part should be of a darker colour, and have a string or dado in colours to relieve the sight. The walls under the bath enclosure should also be covered in the same manner. A bell-pull should be placed in a convenient position to the bather's hand, so that it can be easily reached.

The heat of the bath water should be about 150° to 170°; the quantity required for each bather is about 25 to 30 gallons of hot water, and 8 to 10 gallons of cold water. The cold water should be turned on first, and the hot water let in gradually; this prevents any sudden expansion of the bath, which often causes it to leak at the joints; it also saves the enamel from cracking and scaling off.

The fitting of baths in a proper manner is of great importance. It is advisable to place the work in the hands of firms especially experienced in these matters; when it is done by incompetent men it becomes expensive to maintain it in proper order and repair, and a constant source of annoyance.

The bath room may either be heated by an open fire, or by heating-coil from the warming apparatus, when the house is provided with one; the latter plan gives the most equal temperature in the room, and saves much trouble and expense.

SINKS.

The proper fitting up of these conveniences in a house is a matter of great importance, especially with regard to the health of the inmates; the neglect of proper care may be the means of causing sickness. As a large house will require several sinks in different parts, a few types suitable to their purpose will now be described.

Housemaids' Sinks for the chamber floors should be made of enamelled slate, with a hinged metal grid for pails, &c., to stand on. The waste water should be taken away by a lead pipe 2½ inch diameter, 8 lb. per foot, and carried through the outer wall and discharged over a R.W.P. head, and be fitted at the end with a copper plate flap, in the same way as described for former waste-pipes. The pipe must have a patent solid-drawn lead trap, provided with a cleaning screw placed directly under the basin. No sharp bends should be made in the pipe at any part. At one side of the sink a slate drainage slab, grooved at the top, should be provided.

Hot and Cold Water should be laid on at the basin, and a cold water tap provided, to stand about 14 inches above the floor level, for filling cans, jugs, &c. A lead safing should be provided under the whole of the sink and water-taps before described. The under part of the sink may be enclosed in woodwork, and made to open in front. A permanent ventilator should be placed in the outer wall to ensure constant ventilation. It is advisable to line the walls with plain white tiles, at any rate about half way up the walls. The room should be well lighted, and where the extra cost need not be considered, the floor should be covered with tiles.

Butler's Sink.—This should be made of tinned copper, say 3 lb. per super foot.. The waste-pipe should be lead, 2 inch diameter, fitted with a patent solid-drawn lead trap provided with a cleaning screw; it should be placed directly under the sink, carried through the outer wall, and be discharged over a waste-water trapped gully, described hereafter. Hot and cold water should be laid on. A draining board, covered with tinned copper, should be provided, and racks for glasses, decanters, and plates. The whole may be enclosed in wood panelling and fitted with a hinged cover. Under the sink, taps for drawing water into cans, &c., should be provided. A lead safing, 6 lb. per foot, should cover the whole area of the enclosure, and 1½ in. diameter lead waste-pipe should be carried through the wall and discharged over the trapped gully before named. At the sink the walls should be covered with glazed tiles; they should also be used under the sink within the panelled enclosure. A ventilator should be placed in the outer wall over the sink, to take away vapour.

In large establishments it may be advisable to provide one of George Jennings' patent butlers' sinks : they are fitted up in a very complete manner, and are very efficient and satisfactory in working, they are provided with a special waste valve that prevents silver, &c., being lost down the pipe.

Scullery Sinks.— These should be made of enamelled slate about 1 inch thick. The waste water should be taken away by a 2½ in. diameter lead pipe, 8 lb. per foot ; a patent solid-drawn lead trap fitted with a cleaning screw should be fixed directly under the sink outlet, the pipe should be taken through the wall and discharged *over* a grease trap. Hot, cold, and rain water pipes and cocks must be provided at the sink. A lavatory basin should be fixed at one side of the sink, and at the other side a draining board covered with copper, this is usually preferred to slate, as china is not so likely to be broken on it. The water from the basin must be discharged over a waste-water gully described under head of "Drainage." At the back and side walls of the sink glazed tiles should be used. A ventilator should be fixed in the outer wall directly over the sink. Efficient fresh-air inlets must be provided in the room.

Stone Sinks should not be used. When close economy must be studied, glazed earthenware or fireclay may be substituted; they are very suitable, cleanly, and durable, as well as economical. Bell traps in the sink should *never be allowed*, nor any *movable grating*. The former are a delusion and a snare, and have too long been the curse of households; they ought to be called death traps. There should never be any chance given for opening a communication between the house and any part of the drains or the pipes leading thereto. In cases where there is any unavoidable long length of dis-charge-pipe, it should be provided near the lead trap, with 2 inch diameter ventilating pipe, carried up the house for ventilation. The interior of the pipe itself gets foul, and sometimes causes some slight nuisance when not well ventilated.

CHAPTER III.

DRAINAGE.

THE drainage of a house is a most important matter, and although so essential, it is too often very much neglected. The system to be adopted for a detached house will first be described. The following principles, as before named at p. 5, must be clearly understood :—

1. All the drain pipes or sewers must be *absolutely* cut off from any communication with the house.

2. The soil and kitchen drainage should go into one system, the bath or lavatory water into another, and the rain water into another.

3. All the drain pipes must be kept *outside* the house ; it is advisable to have those close to the house made of cast iron, with socket joints run with lead. The author adopts a special kind, which he has designed to give ready access to all parts.

SOIL DRAINS.

The soil pipes should deliver into special cast iron traps or receivers, with air inlets ; the author uses those of his own design, they are made by George Waller & Co., London. In cases where the soil pipe is above say 20 ft. long, a ventilator pipe from the trap should be carried above the roof, this in *addition* to the ventilating pipe described for the top of the highest point of discharge into the pipe. Air *inlet* pipes must be provided and carried up clear of all windows, in case of any eddy current causing a discharge of foul air.

The end of the lead soil pipe at the point of discharge into the iron receiver should have a gun-metal ring or spigot slipped on and soldered, this is for the purpose of caulking it into the inlet socket of receiver. The depth of water trap or seal between the soil pipe and line of sewer pipes must not be less than 3 inches to 4 inches.

The pipes leading away to the sewer or to the cesspool should be 6 ft. or 9 ft. lengths of 6 in. diameter cast iron pipes, with joints run with lead. They should be laid upon a bed of Portland cement concrete 9 inches thick, and to a regular fall of $\frac{3}{4}$ to 1 inch in each 10 ft.

At all junctions, Y pipes should be used, with examination holes and sweeping eyes ; junctions at right angles should be avoided, but where this cannot be done, the T pieces should be made with long curved junctions, and an inspection hole and air-tight cover provided at these points. The author uses pipes of his own design, wherein special arrangements are made for cleaning, examination, and repairs.

Where stoneware pipes have to be used the joints must be made with Portland cement ; the spigot ends of each pipe must rest upon a bed of cement, and the jointing must perfectly fill the sockets; the inside of the pipes must be perfectly free from any obstruction, especially of any roughness at the joints; they must be laid upon a bed of concrete as before described, in this case ample means of inspection and cleaning must also be provided. The architect should state in his specification that these pipes are to be best tested, and state the maker's name; there is much difference in the quality.

Grease Trap.—The sink outlets discharge over the trap; the author uses a special kind of his own design for this purpose; a movable tray is fitted to the interior for the purpose of taking out the grease and dirt periodically. The outlet pipe from the sink delivers into the open air *above* the grating of the trap or receiver, and *not under* it.

The outlet from the receiver to the drain is connected to a cast iron socket pipe, and the joint made with lead, the junction to the main pipe should be made by an easy curve.

At the highest point of the drain pipes, near the sink outlet, a $3\frac{1}{2}$ in. diam. cast iron ventilating pipe should be carried up above the windows of the house, and an air inlet pipe fitted with a talc valve should also be provided in a suitable position.

Receivers.—In large houses, where more than one soil pipe discharges at one spot, a receiver should be formed of brickwork, lined with glazed bricks, say 2 ft. 6 in. by 2 ft. by 2 ft. deep. The bottom should. be paved with George Jennings' patent vitrified stoneware blocks in suitable sections, having open channels sunk in them to suit the various junctions, the outlet to the drain should be by a trapped pipe, provided with a sweeping eye in the direction of the drain, the end of this pipe must be fitted with an air-tight cover or stopper.

A ventilating pipe 3½ inch diameter of cast iron should be connected with this receiver, and carried to the top of the house free of all windows; the pipe must be close jointed, the top should be fitted with a Banner's patent cowl.

An air inlet pipe and talc valve should also be provided and connected to the chamber, care being taken that it is carried up sufficiently high to be clear of all windows. The chamber is provided with a cast-iron frame and an air-tight cover for examination and cleaning, this should not be less than 15 to 16 inches diameter.

In cases where the drain pipes have to run a long distance before falling into the cesspool, a dis-connecting chamber should be provided, constructed in the same way as the last, and either ventilated by an open grating, or by ventilating pipe and air inlet as before described. The outlet from this chamber should be made by syphon and fitted with a sweeping pipe with air-tight cover as before named, an air-tight manhole must be provided if the re-ceiver is near the house, and must be kept closed.

Inspection Boxes of cast iron should be provided at every 10 to 15 feet; they should have air-tight covers, and be placed at the level of the ground, a length of cast-iron pipe should be attached to the bottom of the box, the other end should be close jointed to the drain pipe, and connected to it by a Y pipe set in the direction of the *fall* of the drain.

All parts of the system of pipes should be readily accessible, each inlet to the drain should be well trapped, and at the highest points well ventilated, and provided with ample fresh-air inlets. At the highest part of the drains a self-acting flushing cistern, containing at least 10 to 15 gallons, should be provided, by the periodical discharge of which the drains can be kept clean.

No other drainage or wastes should be connected to this system on any account.

The cesspool should be kept as far as possible from the house, the dimensions must depend upon the size of house : it should have a cast-iron frame fitted with an air-tight cover to permit of examination and cleaning. In large mansions a special arrangement is needful for the sewage disposal and deodorisation.

BATHS AND LAVATORY WASTE WATER.

This should be made to discharge over R.W.P. heads, provided in each case with a shoe at the bottom, and discharging over improved trapped cast-iron gullies ; the author uses his own design. The gratings are made free in area, to take any sudden rush of water, the outlet is connected either to 4-inch diameter cast-iron socket pipes jointed with lead, or to stoneware pipes jointed with Portland cement. The water should be taken to a soak-away, which should be provided with an air-inlet and also with a ventilating pipe. When the waste-water down pipes are above 10 feet long, a ventilating pipe should be carried from the highest point of the discharge to the roof, the air inlet in this case is through its own pipe at the open shoe at the bottom. The top of the soak-away should have an air-tight inlet cover, to permit of examination and cleaning when necessary. In some soils soapy water causes some trouble, and does not get away very freely, it may be necessary in cases like this to pump out the water occasionally, and run it on to the land or garden.

Town Houses.—Where all the drainage, soil, waste water, and rain water, have to be taken into one system of sewers, the overflow pipes from the waste-water traps may run into the disconnecting chamber at the soil outlets; this will help to cleanse this part of the pipes, while the waste-water discharge will still be disconnected from the main drain, or soil discharge pipes, &c. In any case, when any drain pipe has to be taken from the back to the front of the house *under the floor*, they should be 6-inch cast-iron socket pipes run with lead, and be encased in a bed of Portland cement concrete at least 6 inches thick. A fall of not less than 1 inch for every 10 feet should be given to the pipes. Cleaning eyes should be provided, with air-tight covers, all junctions made by Y pipes and easy curves; no T pipes must be used. The drainage from the house to the main drains can still be entirely disconnected by the system described for detached houses, and the same attention must be paid to the ventilation of the drain pipes. All the other details, as far as the pipes and fittings are concerned, do not much differ from those before described.

RAIN-WATER PIPES.

Rain water from the roofs, &c., should be discharged *over* trapped gullies, as described for bath and lavatory waste water, at p. 26 ; the pipes should have shoes at the bottom, they should be close jointed with iron cement, they should never be directly connected with any drain.

The rain water may be stored in underground tanks and pumped up as required. Some portion can be stored, say 200 gallons, in wrought-iron tanks fixed at a height to command the sink. The tank underground may be made of Portland cement concrete, and the inside rendered with cement, it should be ventilated, and be provided with a manhole to give access for cleaning out.

The provision for the storing of rain water does not receive the attention it should, especially in places where the water is hard and not suited for washing purposes. In cases where only a part of the rain water can be stored, the surplus should be run into the *waste water* drains, and thence to the soak-aways described in p. 27.

LAUNDRIES.

Where one is attached to the house, the waste water can go into the bath waste system of drainage, it should *never* be carried into the soil and kitchen drains ; the waste water should be delivered over trapped gullies in the same way as the bath and lavatory wastes, except that the gullies must be extra large at the gratings, to save any splashing.

STABLE DRAINAGE.

The drainage from the stables and waste water from coach-house and yard should be taken into a separate cesspool ; the pipes may be 6 inches diameter, stoneware, jointed with cement. The yard gullies should be fitted with large dirt boxes, and be well trapped ; those made for the author by Messrs. G. Waller & Co. are very efficient. As the contents of the cesspool from the stables is of manurial value, it should be periodically pumped out into a close vessel by a vacuum apparatus and put on the land. This apparatus is made for the author by the above named firm, and is highly to be recommended for the purpose ; the cesspool can he emptied without the slightest nuisance, mess, or any inconvenience.

The cesspool should be ventilated by $3\frac{1}{2}$ inch diameter cast-iron pipe and be provided with an air inlet pipe 3 inches diameter, fitted with talc valve. When the cesspool is placed at some distance from the stables, a disconnecting chamber or catch pit, as before described, should be formed, and well ventilated.

The system most adopted at the present time for the interior drainage of stables is to make the pavement of the *front* of the stall rather *lower* than the *back;* and in the centre of the width and length of the stall to sink the floor, allowing the drainage to fall into a cast-iron trapped gully provided with a dirt box. The gullies are connected to 6 inch cast-iron pipes, with arrangements for cleaning out when required. The pipes run under the centre of each stall transversely. It has been found by experience that, especially in the case of draught horses, they rest more easily when the floor is constructed in this way. The author is indebted for the suggestion of the level of the floor of the stables to a leading London brewer.

The waste washing water from the coach-house can be taken to a "soak-away." The rain water from the roofs should be stored in a tank placed, if possible, about 6 feet above the ground line; it is most useful for washing carriages, &c.

The drainage of the water-closet for the stable yard and for garden, can be carried into the cesspool for the stables. Pedestal or Vase earthenware closets with flushing cistern and automatic arrangements for periodical discharge are the most suitable and cleanly. The soil pipe should be ventilated and be provided with fresh air inlet as described for the water-closets of the house.

Sewage Disposal.

Sewage Tanks.—In large country mansions away from all systems of sewers, it becomes necessary to collect the sewage in tanks and deodorise it, unless it can conveniently be distributed over a large area of land. The following plan is a simple one to adopt, and is quite efficient for all ordinary cases.

The sewage should pass into a receiver about 3 feet by 3 feet ; in this is placed a basket of open ironwork. All the solid matter and refuse is collected in this. After having been strained it passes out into another chamber about 3 feet wide, and through two filters, one of coarse gravel and one of fine gravel. It then passes into a chamber 3 feet wide, where it is treated with lime water ; then under a dip plate of slate over a weir to the settling chamber ; in this the sludge is deposited, and the top water run off by means of a floating pipe into a small chamber at the end of the last one, and then passes away to any convenient brook or watercourse. The sewage is well purified, clear, and innocuous. The tanks should be made in duplicate, to permit of cleaning out one set at a time ; they should be placed side by side, each set being not more than 6 feet to 7 feet deep, the width depending upon the quantity of sewage to be dealt with in twenty-four hours. Each should hold the whole quantity to be disposed of for 1½ days to 2 days at the outside. The tanks can be covered over by brickwork arches, or by close jointed iron plates, manhole frames and air-tight doors being provided to allow for cleaning out.

Many kinds of chemicals have been used to deodorise sewage, but lime is the most simple, cheap, and efficient purifier.

The sludge deposited in the last receiver or the settling tank is taken out and put on the land. In cases where there is no convenient brook or watercourse near, but where some extent of land is available, the purified sewage may be pumped out into the closed apparatus on wheels described on p. 29, and put on the land as desired. The first strainer is lifted out from the chamber by means of a crab motion when necessary. As before stated, the house and the stable drainage should be kept entirely separate, no connection of any kind should ever be made with it.

Smaller quantities of sewage, that may be run into a stream near, can be treated on a smaller scale in the following manner. At the end of the drain pipes a large catch-pit or receiver of brickwork, say 3 feet diameter, may be formed. In this receiver a wrought-iron open basket may be sunk as before, to receive the solid matters, or they may be stopped by a grid in a small outer chamber, the liquid sewage running through into the 3 feet pit, where an automatic supply of deodorising material may be run in, the overflow to the stream being placed about 36 inches above the bottom, regulated by a weir valve, which is closed while the sewage is in the process of being disinfected. The sludge can be removed as required from this lower part, only the deodorised sewage passing away at the top.

This receiver should be well ventilated, and provided with a fresh-air inlet pipe, be covered over, and have a man-hole frame and air-tight cover.

The sewage in any case should be kept as short a period as possible in the tanks, as it is only when there is time for it to ferment that any deleterious gases are formed prejudicial to health.

The bath and lavatory water being taken into a soak-away, named at p. 27, and the rain water into a separate system of pipes to the storage tanks, provision in the sewage tanks has thus only to be made for the soil and kitchen drainage.

HOUSE-REFUSE, ASHES, &c.

The ashes and dust should be put into a covered receiver made of iron; it should not be too large, and should be cleared once per week. No vegetable matter or other refuse should be put in this receptacle, but all that cannot be burned should be taken out of the house into the garden each day.

In the case of town houses, it is advisable to have two small iron covered bins, one for the dust and one for the refuse that cannot be burned, they should be cleared once or twice per week. Large dust bins should *never* be provided, they should not be placed near the house when it can be possibly avoided. The bins made by Messrs. G. Waller & Co. have been found by the author to be efficient and to answer the purpose; they are well designed, strongly made, and do not get out of order. The old wooden dust bin should be

D

abolished ; this old sinner is responsible for much sickness through the filthy state in which it was usually kept. In many London houses the author has constantly seen it placed close under the kitchen window, and in several instances near the outside *meat* safe.

In large houses and mansions, it is not a good plan to use a dust and ashes shoot from the upper floor, as they are only a lengthening of the dust bin and all its attendant horrors. The dust should be carried down stairs in closed metal boxes, or in places where lifts are used, it can be placed in metal receivers on wheels and can then be run away from the rooms on to the lift. The disposal of house refuse may seem a small matter to touch upon, but it is a very necessary detail to have attention, seeing how much a proper provision for its disposal adds to the comfort and health of the inmates. In the author's opinion, in large towns the dust and refuse should be removed every day from the houses ; this system is carried out with much success in many large places, it should be universal everywhere.

CHAPTER IV.

WATER SUPPLY.

THE supply of water for household and drinking
purposes is a most important matter, and should
at all times receive most careful attention. In
every house two systems should be adopted—one
for drinking, cooking, and washing purposes, and
a separate supply for water-closets.

Assuming that the water in most cases can be
procured from the town water-works, and be delivered at the ordinary height of a dwelling house,
the tanks for storage, and the service pipes and
fittings, will be principally treated. Details are
given, at the end of the chapter, of Wells, Pumping
Machinery, and Motors for driving them, as such
apparatus are usually required in country mansions.

WATER TANKS FOR DRINKING PURPOSES.

These should be placed in the roof, so as to
command all parts of the house. Galvanised
wrought iron is a suitable material for the purpose,
the thickness should not be less than $\frac{3}{16}$ inch;
they should be made rectangular, and rounded at
the corners. At the top of the side plates a ring
of ∟-iron should be riveted on to stiffen them.
A wood cover made with close boards, grooved
and tongued, should be bolted to the ∟-iron,

one part being hinged to give access for cleaning
out. The down service-pipe may either be
of lead or wrought iron; for an average sized
house, say of 14 to 18 rooms, it may be 1¼ inch
diameter, it should stand up about 3 inches from
the bottom of tank, so as not to draw off any
deposit. A perforated rose of gun-metal should
cover the outlet pipe. The rising or supply main
should be carried up inside the house and deliver
over the top of the tank; it should be fitted with a
ball valve to control the supply. A wrought iron
overflow pipe 1½ inch diameter should be fixed in
the side of the tank about 3 inches from the top,
this should be carried through the outer wall of the
house and discharge in the open air where it can
be seen. A trumpet overflow and waste pipe,
washer, and hollow plug, should be fixed at one
corner of the tank; this should come into action if
the former pipe fails. It is also for use when
cleaning out the tank, as all the water can be run
out through it. The waste pipe should be made to
discharge over a R.W.P. head; it should never be
connected with any other waste water pipe, but
always be discharged into the open air. As a rule,
only one down supply pipe is necessary, except
that for the hot water service, as branches from
the main pipe can be carried where required. Light
in the roof should be provided near the tank. The
contents of the tank should not be more than one
to one and a half day's supply, especially where
there is a constant service, it being essential that
the water should be used as fresh as possible.

Lead-lined wood tanks should not be adopted. Cast iron tanks are as suitable as galvanised wrought iron, but are much heavier; unless the former can be put on to strong walls and carried on wrought iron girders, the latter should at all times be used. Tanks should be cleaned out at least eight times per year; a rule to this effect should be posted in the tank room, tenants should have impressed upon their minds the necessity for the strict observance of this rule. It is astonishing how much such a simple and necessary proceeding is neglected, it is often the main cause of serious illness in a household.

No supply for any purposes than for drinking, cooking, bath, hot water service, or washing purposes, should be taken from this tank.

Tank for Water Closet Supply.

The tank should be galvanised wrought iron $\frac{3}{16}$ inch thick, fitted up in the same manner as the last, and provided with a wood cover. It should be placed in the roof, and the overflow and waste pipes treated in the same manner. A separate supply pipe should be carried to the water closets; no connection must be made with this pipe for any other purpose. The down pipe must not be less than $1\frac{1}{4}$ inch to $1\frac{1}{2}$ inch diameter, in order to afford an efficient supply to thoroughly flush the closet, as named at p. 10. In some instances, where Pedestal or Vase closets are used, a separate small cistern is fixed in the room of the closet, they usually are made to contain 3 gallons; the connection from this cistern to the closet-pan should be a 1½ inch. diameter lead pipe; the whole quantity is discharged each time the handle is pulled. In the case of Patent Valve closets, the pipe connections are taken direct from the upper tank to each closet by suitable branch pipes. The inlet valves and regulators for the pans form part of the apparatus attached to them.

PIPES.

Both rising mains and down supply pipes should be run in channels specially formed in the walls, they should never be built in or sunk in the plaster; wood covers should be provided screwed to the side-boards of the channels, with suitable doors at intervals to permit of examination and repairs. The cold water pipes should never be placed in the same channel as the hot water pipes; there is no objection to the down pipe for the water closets being placed in the cold water pipe channel. Each pipe should be painted a different colour, for the purpose of identifying them, say, rising main, *white;* drinking down supply, *blue.* All waste pipes, if near any supply pipes, should be painted *black.* Cocks to shut off should be placed in the rising main inside the house on the ground floor, and one directly under each of the tanks, to close in case of necessity.

HOT WATER SUPPLY.

Hot water should be carried to the principal parts of the house, and should be at all times available at the baths, lavatories, housemaid's, butler's, and kitchen sinks. In some large establishments it is also carried to the principal bed rooms.

The water should be heated in a high pressure wrought iron boiler placed at the back of the kitchen fire. The boiler should be obtained from a good maker, be of ample size, and carefully set ; it should be equal to a pressure of 30 to 35 lb. per square inch. In most cases the hot water circulating cylinder is most conveniently placed a little above the kitchen floor; this is a very safe plan, as, in case of any failure of supply from the cold water tank, all the water can never be drawn off the boiler. The contents of the cylinder should be from 50 to 60 gallons ; it may be circular in shape and about 1 foot 10 inches to 2 feet diameter by 3 feet to 3 feet 6 inches high ; the thickness should be at least $\frac{3}{16}$ inch, the top and bottom cup-shaped and $\frac{1}{4}$ inch thick. The supply for cold water should be taken from the upper tank by a wrought iron pipe $1\frac{1}{4}$ inch diameter; the connection between the circulating cylinder and the boiler should be $1\frac{1}{2}$ inch diameter, and placed at the centre of the cylinder. All the hot water is taken from the cylinder, the rising pipe being at the top ; the first branch will be for the scullery and kitchen purposes, also the lavatories and pantries situated at the basement or ground floor level. The second will probably be for the baths and housemaid's room on the first or second floor, as the case may be. When the highest service is taken off, the "return" or down pipe is taken to the circulating tank and connected at the bottom ;

this pipe may be 1¼ inch diameter. At the highest point of the " flow " pipe, an expansion or relief pipe is provided ; this should be passed up through the roof. Hot water may be drawn off at any point of the supply or " flow " pipe, but no connection should be made to the " return " pipe. The supply of water is kept constant from the cold water tank ; it enters through the boiler and cylinder at the bottom. A safety valve is provided at the top of the circulating cylinder, to give relief in case the expansion pipe should not act. A cock is fixed at the bottom of the cylinder to empty all the water out of the boiler and system of pipes when necessary. A cock should be fitted in the cold water supply pipe at the bottom of cold water tank to shut off the water in case of necessity. The connections to the baths and lavatories are made in the same way as for ordinary cold water service. The size of the supply pipes to the lavatories and sinks should be ¾ inch diameter, and to the baths 1 inch diameter.

The pipes should be felted and covered with canvas, and run side by side in channels in the wall, the " flow " or supply pipe having the canvas covering painted *red* and the " return " pipe *green*. The horizontal pipes must either be kept level or on a slight rise to the apparatus to be supplied ; no dip must be made in any of the pipes, or the circulation of the water in the system will not be perfect. No water for heating coils should be taken off any of these pipes.

In Mansions and large establishments where

there are several baths and lavatories, and a large quantity of hot water required, a separate boiler must be provided; for moderate sized places a wrought iron saddle boiler is sufficient, but in very large places one of the Cornish or plain cylindrical kind is the most suitable.   This should be placed at the lowest level of the house, well below all points where a supply of hot water is required.   When there is a large demand for hot water, the circulating tank should contain one hour's supply.   The cylinder is to be placed on the same level as the bottom of boiler, and is fitted up in the same way as described for the smaller hot water circulations worked by the kitchen boiler.   The "flow" and "return" pipes in these cases should be 2 inches to 2½ inches diameter, they may be made of wrought iron screwed together.   The boilers are set in brickwork; the furnaces are made large, in order to burn small fuel and refuse.   The greatest care must be taken in fixing the pipes; when placed horizonally, they must have a *gradual* rise from the supply pipe—the pipes must not be *dipped* at any part.   Proper expansion joints must be provided, and the pipes left free to move as they expand and contract.   Air-cocks should be placed at certain points in the mains; these must be opened occasionally, to keep the pipes free from air, and to ensure a perfect circulation.   An expansion pipe must be provided at the highest level of the flow pipe and carried through the roof to the outer air, standing up about 14 inches, and bent over at the top.

WELLS AND PUMPING MACHINERY.

In large country houses and mansions the water supply is usually obtained from wells, and is pumped up by hand or steam-power to tanks placed in the house, some at the top under the roof to command all the rooms, and some at lower levels for kitchen and other uses. No more water need be pumped to the upper tanks than is likely to be used, as it would be waste of power and fuel.

Wells.—The position of the well should be carefully chosen ; be sure the surrounding ground is not in any way contaminated with any cesspool drainage, or near any adjoining property where any other kind of deleterious matter is likely to be in the earth. *It must never be placed near any cesspool.*

It would be imposible in a short treatise of this kind to enter much into the subject of wells, as it is far too extensive a matter ; assuming, however, instances where the water can be obtained within say 70 or 80 feet of the surface, the following may form a guide for such simple cases.

If the well passes through gravel or sand, it should be lined with brickwork 9 inches thick, set in Portland cement ; the top or surface waters should be shut out. When the gravel is porous, and contains much surface water, it is necessary in most cases to shut it out by sinking cast-iron cylinders. If at the lower level the well passes through hard chalk, this portion need not be lined with brickwork. If, at a depth of 90 feet to 100 feet from the surface, water of a suitable character is not met with, a boring should be made, and tubed until the spring is reached. Anyone wishing for further details as to wells is referred to the author's book on "Breweries and Maltings" : E. and F. N. Spon, London.

Pumps.—These must be sunk in the wells to a depth allowing the suction valves to be within say 20 feet of the lowest permanent water line. The kind of pumps to be used will depend upon the quantity of water wanted ; as a rule, a set of three throw lift pumps requires one suction pipe and one rising main. The pumps may either be worked by hand-power, horse gear, or by the steam motor hereafter named. Windmills have also been applied to work pumps with much advantage. In cases were single pumps for small quantities of water are used for deep wells, the pump rod should be balanced by a lever and weight.

Steam Motor or Engine.—A very suitable and simple machine for working pumps is Davey's patent steam motor. It is perfectly safe, very economical as to fuel, and can be worked by any domestic. There is no danger from any explosion ; the fire only requires attention three to four times per day ; coke or cinders will keep it going. It takes very little room, and can be placed at the side of the well in a small house. Anyone requiring more detailed information as to this engine, or as to pumping machinery of any kind, is referred to the author's book upon "Pumps and Pumping Machinery," parts 1 and 2 : E. and F. N. Spon, London.

Hydraulic Rams.—In some instances water can be obtained by gravitation from the surrounding hills; in this case it may be conducted by pipes to a small covered reservoir or receiver, placed at a level to work a hydraulic ram, which would force the water up to the level required in the house. The size of this machine depends upon the quantity of water required, and the height to be raised above the ram ; it is automatic in its action, requires very little attention, and may be kept working night and day if desired ; and does not want many repairs.

RAIN WATER.

In treating of water supply, the author must again allude to the necessity for storing the rain water for use.   To determine the size of the tanks, take the area of the roofs and allow 7 inches deep for two months' rainfall ; this will give enough for any house supply under ordinary circumstances. The tanks may be made of cement concrete, and rendered inside ; they may be placed underground and covered over by iron plates close jointed. The tanks should be ventilated.   A manhole and cover should be fixed at the top to give access for cleaning out.   An overflow pipe should be provided near the top, to discharge over a trapped gully sunk at the side of the tank, the water from which discharges into the waste water system of drainage.

The water may be raised from the underground tank to the level of any upper tank by a hand pump; this may be fixed on a board in any convenient position. The *length* of the suction is not much importance, but the depth from the suction valve of the pump to the bottom of the tank must not exceed about 25 feet. A rose and strainer should be fixed in the suction pipe, to obtain the water clear and free from dirt. The drain pipe leading the water to the tank may be fitted at the end with a dirt-box and filter to prevent the dirt and grit from entering the tank.

CHAPTER V.

WARMING APPARATUS.

In houses of average size, and in mansions, it is very essential for comfort to warm the halls, corridors, staircases, bath-rooms, &c. There are two systems, "high pressure" and "low pressure" : the former plan has been perfected by Messrs. Perkins and Sons, and is very economical and easy to work ; it is somewhat more costly as to first out-lay than the low pressure system ; in many cases, however, where first expense is no object, the high pressure system may be the more suitable. It is advisable to seek the advice of those who have had special experience in this kind of work, and thus be able to decide upon the best system to adopt, and to make provision in the building during construction. Both systems will now be described, and full details given.

High Pressure System.

The author has found Messrs. A. M. Perkins and Sons' the best ; he has had places heated in a most satisfactory way by this firm.

The apparatus consists of a boiler or furnace containing a coil of thick pipes $\frac{7}{8}$ inch diameter inside, and $1\frac{5}{16}$ inch diameter outside. A coil is placed at a lower level than the circulating pipes in

a furnace, and is surrounded by the fire ; a wrought
iron box encloses the coil and furnace, varying in
size according to the space to be heated.

From the top of the coil an endless tube is
formed, passing through the various rooms, either as
lines of piping or divided into coils and contained
in ornamental perforated cases. The endless tube
is filled with water and closed at all parts. An
"expansion tube" is placed above the highest part
of the circulating pipes, and is usually of much
larger diameter than the other pipes ; the contents
of this larger tube should not be less than $\frac{1}{12}$ of the
contents of all the pipes.

This tube is for the purpose of allowing for
the expansion of the water as it gets heated. At
the highest part of the coil, but *separate* from the
"expansion" tube, a cock is provided for filling the
apparatus, but arranged so that the *expansion* tube
is left *empty*. A screw plug is sometimes used
instead of a cock.

As the water becomes heated in the coil in the
furnace, it rises from the top of it to the highest
level of the circulating pipes, by this means a
column of water is formed lighter than the colder
water, which descends to the *lowest* part of the
coil in the furnace. Eventually all water in the
pipes becomes heated, and a perfect circulation is
maintained. The fire is regulated by dampers, &c
according to the heat required in the rooms. When
the apparatus is cold the expansion tube is empty.

The tubes are made of wrought iron, and connected by right and left hand screw collars or

sockets, 2 inches long ; the end of one pipe is turned
to a flat face, and the pipe to which it is jointed to
a V shaped face. Red lead or any other jointing
is not required, the solid metallic contact is alone
depended on.

The quantity of pipe required is 1 foot run for
54 cubic feet of space to heat to 65°, about 1 foot
run for each 40 cubic feet space for 54°. The
number of windows in the room and their aspect
must be taken into account and duly allowed for,
and in cases where the room is subject to cold
draughts of air, the quantity must be increased
about $\frac{1}{10}$.

The fuel best suited for the furnace is coke ; it
is fed at the top. The pipes want replenishing
with water once per week ; a very small quantity
is required. The apparatus wants very little
attention, any domestic can see to the fire and
regulation of the dampers. The fuel used is very
small. One great convenience of this apparatus
is, the pipes being of small diameter, occupy little
space, and only small cutting away is needed
to pass the pipes through any wall or floor, they
also lie very closely against the walls, and only
require to be hung upon small hooks. This class
of apparatus is *perfectly* safe, the pipes are tested
to a pressure very considerably higher than they
ever have to bear. The pressure cannot be materially increased by the attendant who works
it. It is cleanly in operation ; the furnace oc-
cupies a small space, and may be placed in any
room where a smoke pipe can be carried away.

For moderate sized jobs, the enclosure of the furnace does not exceed 3 feet by 3 feet ; it stands on iron feet, and can be placed upon a stone or iron plate in the floor. The inside of the iron enclosure of the furnace is lined with fire-brick. As the apparatus is patent, and only erected by special people, no further details are necessary. The system has been largely used, and is very efficient in action.

LOW PRESSURE SYSTEM.

The boiler must be placed at the lowest part of the house, in the basement ; it may be the " Saddle " kind, set in brickwork. The " flow " and " return " pipes may be wrought iron, 2 inches diameter, provided with suitable expansion joints and air cocks at all bends to take away any air. The pipes should have a regular rise from the top of the boiler to their highest point, and a regular fall to the bottom of the boiler.

An expansion pipe, open at the top, must be placed at the highest point of the " flow " pipe, to give relief if the water gets too hot. Air cocks must also be fitted to the pipes. No part of the pipes must be dipped; great care is required as to this.

Coil Boxes or Radiators are the best means to adopt for heating. Pipes sunk in trenches are not suitable in any part of a house; they become receptacles for dust and dirt, and are in other respects objectionable.

The size of the coil boxes or length of the pipes

E

is arrived at by taking the cubic contents of the place to be heated. Halls, passages, staircases, &c., should not exceed 55°. For each 1000 cubic feet of space allow about 14 feet run of 2-inch diameter pipe in the coils ; in places where the windows are very large, or where the situation is very much exposed, allow say $\frac{1}{10}$ extra length of pipe.

Pipes 2 inches diameter are more suitable for house work. The "flow" and "return" pipes may be carried in the same channels as the hot water service ; and should be painted a different colour to distinguish them. The pipes should be provided with expansion joints, and must not be rigidly held at any part. They should rest on rollers, hung in cast-iron chairs or carriages. In large places 3-inch or even 4-inch pipes may be used ; they take longer time to heat up, but do not cool so quickly as 2 inch diameter. (See Heating of Greenhouses, Chapter X.) The supply of cold water to the boiler may be taken from the house tank to a small feed cistern, holding 3 or 4 gallons ; this is provided with a ball valve to control the supply. In some instances it is desirable to have troughs of water on top of the coil boxes and radiators to moisten the air. All the passages, &c., should have good top ventilation near the ceiling, and be provided with fresh air inlets. Rooms that are only occasionally used, such as libraries, are very suitably warmed by coils ; the same applies to bath-rooms and bed-rooms, where a moderate and regular heat is required to be maintained. In entrance halls, coil boxes enclosed in ornamental cases with marble tops make the best appearance and are also very convenient for use as sideboards. The apparatus for heating requires very little attention ; small coke is sufficient to keep it going. The fire can be banked up at night.

Conservatories, when attached to the house, can be heated from the same source. Billiard and smoking rooms can also be heated with coils with much advantage, it only being necessary to turn on the valve a short time before the room is to be used.

Proportion of Boilers.—When the total length of pipe has been decided, select a boiler about 25 per cent. more power than stated by the makers, to ensure of a good circulation of hot water without any forcing of the fire. In addition to the saddle boilers mentioned, there are several others equally suitable, such as small vertical boilers with quick circulation ; also sectional boilers. These latter, however, are more complicated, and not so suitable for ordinary purposes, especially where there is any deposit in the water. Boilers should be as simple as possible : complicated cross tubes and chambers are of very little advantage, and are more liable to silt up, and then soon burn out.

HOT AIR APPARATUS.

This system is sometimes used in large houses ; it is not, however, in the author's opinion, so suitable as hot water. It is more expensive in first cost, not so safe as regards fire risks, and not so easy to

control.   There are many instances where it may
no doubt be used with advantage, such as in
churches, public rooms, &c.   It cannot be recommended for house warming.   One objection to it
is, the air is too dry, and it often happens that dust
gets heated and carried into the rooms with the
air ; this causes a very unpleasant irritation of the
skin and throat.   This difficulty has been modified
by passing the warm air over shallow dishes of
water; this is a great improvement, as the air is
moistened and more pleasant to breathe.

STEAM HEATING.

Only public establishments, as a rule, are heated
on this system.   The arrangements do not materially differ from the low-pressure apparatus, as far
as the pipes, coil cases, and radiators are concerned.
The heating power is taken from any convenient
boiler ; any pressure above 10 lb. per square inch
will do.   It is not so convenient as hot water
for heating residences, and is more likely to give
trouble by leaks in the joints of the pipes, unless
much care is used in turning on the steam.   The
condensed water formed in the pipes is taken away
by condense boxes, the overflow from which is
taken to any convenient drain.   The quantity of
pipe required is not much less than for the hot-
water low-pressure system, except in cases where
higher pressures of steam are used than named
above.   When rooms are heated by steam, ventilation must have careful attention, both for the inlet
of fresh air and the outlet of moist and foul air.

CHAPTER VI.

GAS AND ELECTRIC LIGHTING OF HOUSES.
ELECTRIC BELLS AND TELEPHONES.

GAS LIGHTING.

THE proper system for laying on the pipes and connections seldom has the attention it deserves. A few suggestions on this subject may be of some value. It is not advisable to leave these matters entirely in the hands of gasfitters ; it is far better for the architect to state what he wants done in sufficient detail to ensure that the general system of the work is correct, and pipes of proper size used.

The Meter should be placed at the lowest part of the house in a cool position, if possible near a window, where it can be easily seen. The tenant's fittings commence the house side of the meter. All the main and small service pipes should be wrought iron ; no composition pipes should be used. The main service pipe for a house requiring say forty to fifty lights should not be less than 1 inch diameter ; this should pass up inside the house, reducing to ¾ inch and ½ inch at the chamber floors. At each floor from the main, cross service pipes should be connected by "cross-pieces," one end being fitted with a screw plug, to enable the pipe to be cleaned when any deposit forms. No part of the small service pipes should be less than $\frac{3}{8}$ inch diameter ; the down pipes to the chandeliers and brackets should be connected by "cross-pieces." At every 9 feet to 10 feet the pipes should be fitted with a "disconnector," to enable them to be easily taken down for examination, cleaning, or any alterations. At the lowest part of the main a syphon and draw-off cock should be placed, to keep the pipes free from water and other deposit. If possible, the vertical gas pipes should be put into channels in the walls, but *not* in the same channel as the water pipes ; they should never be built in the walls or plaster. The horizontal pipes to the floors should be placed between the joists, the floor boards over them being screwed down to allow for easy removal.

The outside pipe from the company's mains should not be less than $1\frac{1}{2}$ inch diameter ; it should be well protected from the weather, and be fitted with a stopcock near the road, to shut off in case of need. A syphon box should also be placed on the line of pipe to catch any water or deposit. It must be borne in mind, when the pipes are exposed to the weather, the napthaline and water in the main is occasionally frozen, thus causing a stoppage of the supply.

Only first class firms should be employed to do this kind of work. The author has found in country places, except those near any of the large manu-facturing towns, great difficulty in getting the work well done. He has often proved it to be the cheapest and best plan to put the work into the hands of a London firm, who not only do it much better, but as a rule at a less price. Bad gas-fitting is often one of the greatest troubles an architect has to deal with; it leads not only to annoyance, but in too many cases the leakage arising from careless work proves most injurious to health, and the cause of explosions and other sad accidents.

Ventilation of the Rooms.—This is a very essential thing in gas-lighting. If the house is being newly built, the following should have attention:—At the ceiling over the chandelier an opening should be left, communicating by means of flat zinc tubes with the outer air, the outlets being so arranged that no back current can enter the tubes. In places where the brackets or branches are near or fixed on the walls, the cornice may be perforated in the same manner as the ceiling. It is absolutely essential that the products of combustion from the gas should be carried away from the room.

An inlet of fresh air should be provided by a " Tobin's " or " Boyle's " patent ventilator; these are fitted with valves, and can be regulated as desired. The tubes should allow the air to enter at about 5 feet from the floor. Where these tubes cannot be used, a "Sheringham's " ventilator may be placed over the door of the room. All parts of the house where gas is used should be ventilated in the same manner. At the top of the staircase, in the roof, a " Boyle's " ventilator should be provided ;

the hot and foul air should always have an outlet
at the highest point of the house. Where heating
coils are used in the halls and passages, the fresh
air may be brought in at these points ; if not, a
"Sheringham's" ventilator should be fixed in the
outer wall of the house. It is useless to make
outlets for the hot and foul air if inlets for fresh air
are not also provided. Particular attention is called
to the ventilation, as it is a most important matter,
and one which is too often neglected.

Gas Stoves may be convenient to use in entrance
halls and other places ; the greatest attention
should be paid to their ventilation, otherwise they
are most unhealthy ; they are not to be recommended for bedrooms, except in the case of
"George's Patent Calorigen," where the gas
burning and its products are hermetically sealed
from the room, the air to be warmed being brought
in through the outer wall ; the products of combustion are taken outside the house, and the air to
supply the stove is brought in from the outside.

Gas Fires may be used for billiard rooms in the
usual grate with asbestos or coke, the latter is to
be prefered ; the burners should be of the Bunsen
kind. The most perfect ventilation must be given,
and all down draught in the chimney avoided.
Some of the best gas stoves are made by Fletcher
& Co. ; they are very economical in working, and
do not cost more than an ordinary open fire, if so
much. A free supply of gas must be given.

ELECTRIC LIGHTING.

In large mansions, where the power cannot be obtained from an electric light and power company, and also where the installation is of some extent, the architect is advised to call in a consulting engineer experienced in such matters, and place the matter in his hands, as it requires a large amount of special experience to carry it out in a satisfactory manner. The author must not be thought interested in this case, as he is sometimes entrusted to carry out such matters jointly with architects. Experience has, however, proved that the services of a specialist is advisable in all such cases. The dynamos and driving machinery, with boilers, and the best system of wiring, are matters that should be drawn up in detail in specification, and only those firms permitted to tender who are manufacturers and who have had experience in such work. The dynamo and boiler rooms require special preparation, which, if done on the advice of an engineer, will save both trouble and expense.

Large Installations.—Where a complete electric light plant has to be laid down, a few general outlines of the machinery necessary may be useful as a guide to the architect. If no boiler and engine power are available on the premises, it will be necessary to provide them.

For small installations a combined engine and boiler will be sufficient. The dynamo may be driven by a strap off a pulley on the engine.

For larger installations the boiler may be made separate, and the engine shaft attached direct to the shaft of the dynamo, forming one machine. This latter is usually the best plan to adopt. In this case the engine and dynamo are fixed on the same bed plate. The speed of the dynamo is from 350 to 500 revolutions per minute. The steam pressure should not be less than 80 lb. per square inch; the stroke of the engine is not more than 6 inches.

There are many forms of dynamos. Those giving continuous currents are the most suitable for "incandescent" lighting. The most approved makers are Siemens Bros. & Co., Crompton & Co., J. H. Holmes & Co., Mather & Platt, and Paterson & Cooper.

A switchboard should be provided for distributing the electric current. The cable or leading wire must be proportioned according to the distance the current has to be carried and the number of lamps.

The power required in the engine is, 1 indicated horse power for each 10 "incandescent" lights of 16 candle power each.

The boiler power, to give sufficient steam, is about 4 nominal horse power for each 100 lights of 16 candle power each. The nominal horse power is calculated on 1 cubic foot of water evaporated per hour for each horse power.

Each indicated horse power of the engine takes about 27 lb. of steam with compound engines, and about 35 lb. in simple engines.

The foundations of the engine and dynamos

should be isolated from the walls; very special precautions are requisite, which require the skilled attention of the consulting engineer to adapt them to each particular case.

All single lights up to 300 candle power are best upon the "incandescent" system. The cost of the "arc" system is much less than the "incandescent," but is only suitable for out-door work. The lights are perfectly safe, cannot be blown out, give a steady light, and require no attention or daily renewal like the carbons of the arc light.

Wires or Conductors.—The temperature should never exceed 150° F.; their size, and the system of fixing, require great attention; the two wires may be carried in the same casing, but must be kept separate.

Safety Fuses should melt and break the circuit when the current increases 50 per cent.; only the best kind should be used; the author believes Hedges' Patent acts in a very satisfactory manner.

Power from Electric Light Company's Cables.— When power can be procured from a company's cable, the "wiring" and other details for lighting is a much less complicated matter. The size of the wires will depend upon the number of lights and the distance or length from the supply cable off which the power is taken. The wires can be carried in a very small space, and through wood boxes or casings, which in many instances may

form part of the woodwork of the room, &c. The
wires have to be well and carefully insulated, and
all the work must be done in accordance with the
rules of the Phœnix Fire Office. The brackets for
lamps and electroliers are made in great variety by
Strode & Co., Faraday & Co., Edmundson & Co.,
Verity & Co., and other equally good makers. Any
of these firms also undertake the "wiring" and
fitting up the whole work. All the lamps, switches,
resistances, and other requisite things should be of
the best quality. Competition is so great in this
line that unscrupulous firms often put in very indif-
ferent materials and work, which always lead to
great trouble when the light is in operation.

As a rough rule to form an estimate upon, the
cost of "wiring," including lamps and plain brackets,
is about 20s. to 25s. per light. The number of
lights required in a room is about the same as for
ordinary gas lights. "Incandescent" lamps of 16
candle power are the most suitable for rooms.
The cost of these is about 4s. to 4s. 6d. each. They
last about 1000 to 1200 hours, or about the usual
lighting time of one year in an average household.
The price charged by the Supply Companies is
equal to gas at about 4s. 2d. to 4s. 6d. per 1000
cubic feet. A 16 candle power electric lamp will
light about 30 square feet. For large rooms, say
27 feet by 18 feet, eight to ten 20 to 25 candle
power Swan lamps are sufficient, and will light the
place well in every part.

Arc Lights are not suitable as a rule for inside

lighting; they may, however, be employed to light the exterior, or any part of the grounds. The carbons last from 12 to 16 hours, according to the strength of the current; these lamps are usually of great power; there are many kinds in use. The light has now been made very steady, by improvements made in the automatic apparatus for regulating the carbons, and also by their superior quality.

For House Lighting, secondary batteries are sometimes employed; they are charged from the supply companies' cables. A battery to operate 25 lamps of 16 candle power to last for 5 hours each night, requires 10 lb. weight of battery for each hour of burning one lamp. Thus $25 \times 5 \times 10 = 1250$ lb. The efficiency of these batteries is from 85 to 90 per cent. of the current used to charge them; up to about 14 days, if the cells are left standing, the loss is about 8 to 10 per cent.

Central Power Stations.—As these places are making great advance in London and some of the large towns, and as it is proposed to bring the power to each house in much the same way as the supply of gas, it enables architects in good houses to make arrangements for putting in the electric current instead of laying on the gas. The difference in the cost of the light is now, gas 2s. 10d., to say 4s. 6d. for the electric light, this price would not deter many from using it. The heat from the electric light is not more than one-sixth of the heat given out by gas. It is very safe to use, especially in bed-rooms, as no accident from fire can possibly take place through its use ; in the event of an incandescent lamp breaking it goes out at once ; no damage can possibly take place under any circumstances.

ELECTRIC BELLS.

These are very superior to the ordinary bells and wires ; in most instances in good houses it is advisable to provide them ; the wires for this purpose should be about No. 20 B.W.G., well insulated.

The battery to work the bells should be Le-clanché's, the number of cells will depend upon the number and position of the bells. For single bells, or when two or three are placed at a short distance, one battery is enough ; the cells are of different powers. Judgment must be used in deciding the power and number necessary for the particular case. The wires are connected to the pushers or buttons, which are placed in the usual positions of ordinary bell-pulls.

An indicator should be fixed in the servants' hall or near any attendant whose duty it is to attend to the bells ; directly a bell is operated in any room it shows upon the board, by a number sliding out, where attention is required. When this is noted, the attendant by touching a button moves the sign and replaces a blank. There are many special arrangements of these useful apparatus suitable for different places and purposes. Any of the firms named for wiring, &c., for electric light, also undertake this class of work.

The wires should be encased, and care should be used to see that they are well insulated; only ordinary care is wanted in using these bells. The batteries require occasional attention to see that they are properly charged and in order; they should be placed in a suitable position, cool, and where they are not subject to variable temperature; they should be enclosed in woodwork to keep them clean, and placed where they cannot be interfered with.

Code of Signals.—By arranging a code, the number of touches on the buttons may be made to indicate certain agreed things for messages that are being constantly sent; this saves much trouble and labour, and in practice works very well.

TELEPHONES.

These may be provided to communicate between the dining room and kitchen, from the hall to the nurseries, stables and other parts; they are only suitable for large establishments, as they are rather expensive. They are worked off batteries in much the same way as electric bells. One very great advantage of the telephone is, a personal communication can be made; the person receiving the message can tell who is speaking and giving the order; there can be no mistake about the authority in the latter case. It is needless to say, much trouble and labour can be saved by their use. The architect must use his discretion as to the necessity for adapting them to his particular case ; and, as the cost is materially increased by the number of mouth pieces, no more should be provided than is absolutely necessary.

CHAPTER VII.

LIFTING MACHINERY.

IN most modern houses of any size where there are more than two floors, or in cases where the kitchens and coal cellars are in the basement, lifting machinery is a very necessary adjunct, and effects a great saving in labour and trouble. For residences of the average size, as well as mansions, lifts may be provided for raising food, also for coals and bringing down ashes, and in some instances for carrying passengers. Each of this class of lifts will be treated in sufficient detail to indicate the best kind to use, and also so as to give an idea of the necessary provisions to be made in the building during construction; a want of knowledge of such requirements often causes increased expense in putting in the machinery on a subsequent occasion. The author advises architects not to place themselves entirely in the hands of lift makers, as it unfortunately happens keen competition is often the cause of very poor and unreliable machines. Added to this, the designer is in a better position when he is able to specify what he requires in some detail; it also places the firms who tender upon a fairer basis, nothing being left to the imagination of any of the manufacturers.

F

Food Lifts.

These may either be made as "single" or
"double" lifts; unless it is a large house, the
former are sufficient and cost less. The cage or
box in which the food is placed is made of woodwork, with one or more shelves to take plates and
dishes. A piece of T-iron forms the lifting bar.
Iron plate rubbing guides are fixed to the box at
top and bottom. The guide bars are, in the best
lifts, made of T-iron, say 2 inches by 2 inches,
these can be fixed to timber runners or direct to
the trimming joist at each floor. At the top of the
framing of lift a spocket wheel is fixed, the rope
attached to the top of the cage passes over this
wheel and is then attached to a cast iron counterbalance, which has grooves planed out on each side
and is guided in L-iron guides at the side of the
cage. On the same spindle carrying the spocket
wheel, a larger wheel is keyed; this is worked by
an endless rope, which passes through eyelet holes
in each floor to the bottom of the house. A small
brake wheel is also keyed on the same spindle, and
is fitted with a brake strap and lever, to which is
attached a small rope, which also passes through
eyelets in the floor to the basement.

The lift hole should be enclosed, and at each floor
a slab should be fixed, 3 feet above the floor-line,
to land things on. A shutter should be provided
at each landing place and also at the bottom. If
these shutters are not kept closed when the lift is

not in use, the shaft becomes objectionable, as the fumes from the kitchen are taken up through the house, and in case of fire it would act as a conductor to carry flames through the place; in some cases the fire insurance companies insist upon the lift holes being enclosed in brickwork.

When these lifts are made "double," one cage is made to balance the other; the counterbalance weight can in this case be dispensed with; in other respects they are constructed in much the same manner as the "single" lifts. A rope is attached to the bottom of each box or cage, and passes over a lower wheel.

Food Lifts for Heavier Weights.—When weights of above ½ cwt. have to be raised at one time, it is necessary to provide spur wheel gear; this requires an extra shaft. The endless working rope wheel is keyed on to a shaft placed above the lifting wheel shaft; a spur pinion is keyed on; this gearing into a spur wheel keyed on the lower shaft. The proportion between these wheels is decided by the load which is to be raised. The power that can be exerted by a female servant on a rope to raise a weight is about 20 to 25 lb.; a man will exert a force of 30 to 35 lb. The details of construction are much the same in other respects as the lift first described. All the parts are made heavier. The cage should be framed in light L and T iron, and lined with wood, the floor being made of oak or beech.

COAL LIFTS.

These are constructed in the same way as the last named lifts, except that the cage is made rather heavier, and no shelves are required. The openings at the floors in these cases may be level with the floor. This is essential where the coal and cinder boxes are placed on wheels ; it lessens the labour, as nothing has to be lifted out of the cage, but can be run off at the level of the floor. The only objection to it is the danger that may arise owing to leaving open the doors. This danger may be partly lessened by providing guard ropes or bars ; but in houses where there are any young children, the opening to the lift should be placed at 3 feet from the ground, in the same manner as the food lifts, and the coal, &c., taken up in hoppers or scuttles ; these may be covered, to save dirt.

HYDRAULIC LIFTS.

Lifts for each of the above purposes may be worked by hydraulic power. In London and some other large towns, water under pressure can be obtained from the mains of the " Hydraulic Power Company " at a very moderate rate. If at a country house, the requisite pressure may be obtained from the house supply tank, provided the pressure is not less than 20 to 25 lb. per square inch ; this means that the tank must be about 57 feet above the bottom of the lift. There are two types of these machines : 1st, direct acting, by using the water pressure directly under a ram working in a cylinder sunk in the ground; and 2nd, those made with a cylinder and movable pulleys or racks and pistons, and large drums for chains to wind on. For lifting food and coals, &c., the latter are mostly used; they are much less costly than ram lifts, and do not require a well or bore hole made.

FOOD AND COAL LIFTS WITH SHORT CYLINDERS.

The cylinders are made about 5 to 6 feet long, and are proportioned to the exact height to be lifted. Thus, if the height is 30 feet, the stroke of the piston of the cylinder would be as 5 to 1; if 50 feet, the stroke may be made 10 to 1. The cages are usually suspended by short linked chains, or steel wire ropes may be used. The cages, guides, and counterbalance are the same as for the hand-power lifts. The grooved lifting wheel at top of the lift should be of the same size as the distance from the centre of the cage to the centre of the counterbalance weight. No small grooved wheels should be used; in any case they must not be less than 9 or 10 inches diameter. The lifting rope should be taken as direct as possible to the cylinder pulleys or winding drum. The counterbalance may be dispensed with whenever the pressure exceeds 50 lb. per square inch.

RAM LIFTS.

These can be used when the height is only one floor, or, say, about 10 feet. A bore hole is made in the ground equal to the height to be raised; in this is sunk a cylinder, having a hollow ram working inside; the cage is fixed direct to the top of the ram ; the guides, cage, &c., are the same as the hand-power lifts, but made rather stronger. A valve-box is provided, and gear rods to work it are taken to the upper floor, so as to control the lift from either floor. There are several modifications of this lift, but the general principle is the same. A very small ram suffices when power can be obtained from the Hydraulic Power Company's mains ; in this case no counterbalance is required.

Passenger Lifts.

For invalids and aged people these are very useful appliances. They should be of the "direct-acting ram" kind; they can be worked by a pressure of water of 20 to 25 lb. per square inch ; the greater the pressure the smaller the ram may be made, and the first cost will also be less.

The ram and cylinder, with valves, &c., are the same as before. When only one floor is to be passed through, a platform railed at the sides is used instead of the cage ; this is fixed directly to the top of the ram. Invalids can be wheeled on to the table or platform, raised to the level required, and wheeled away to their rooms without any exertion on their part; the motion is steady and noiseless. The author has applied these lifts in private houses and worked them from the Water Company's mains.

When required for a large mansion, to travel through several floors, the same plan is adopted ; the ram in this case is put together in sections, and the cylinder made in 9 feet lengths bolted together. When worked off the Hydraulic Power Company's mains no counterbalance is required. As these are much more important machines, the architect is advised to get advice from an engineer before he commences his building, so as to make suitable provision for all parts of the machinery.

There are a large number of lifting machines suitable for special purposes, full details of which will be found in the author's book on " Hydraulic Lifting Machinery ": E. & F. N. Spon, London. As the subject is extensive, it is impossible to do more in this book than indicate the apparatus suitable for most general purposes. In all large buildings, especially blocks of offices and chambers, lifts are being extensively used. The author advises architects only to adopt the best kind and quality, and not to be tempted, by apparent first cheapness, to sacrifice efficiency in working, as well as safety. Lifts can be made to work without any noise ; it is only a question of proper design and good work.

CHAPTER VIII.

GAS AND STEAM COOKING APPARATUS.

In large houses and mansions, in addition to the ordinary cooking range or kitchener, both gas and steam cooking apparatus may be advantageously used. They should be placed in a separate kitchen, well ventilated and partially cut off from the house. A description is given of apparatus suitable principally for large places; many of them may be obtained in smaller sizes. For smaller houses some of the gas apparatus are made more universal, so as to be fit for carrying on several cooking operations at the same time.

Gas Ovens.—The size of these will, of course, depend upon the amount of cooking to be done. A convenient size for a moderate family is 2 feet 6 inches by 2 feet 6 inches by 4 feet high; they should be lined with glazed tiles, and be provided with trays or grids at convenient levels. The gas pipes should be placed in the inside at the bottom of the stove for baking, and on the outside rings of gas should be provided at the top plate to suit various sized saucepans, kettles, &c. This stove will cook about 70 lb. of meat in three hours, at a cost of about $\frac{3}{4}d.$ to $1d.$ per 10 lb. of meat cooked. Pies and puddings may be baked, also bread, cakes, &c.

Gas Frying Plate.—For cooking soles, chops, &c., the size may be 3 feet by 2 feet. The top is provided with round and oval holes and burners to suit various sized cooking vessels. The gas used would cost about 1½*d.* per hour. For stewing purposes these plates are very useful, as the heat is constant when regulated to the proper requirements. They are useful for many domestic purposes.

Gas Broiling or Boiling Stove.—The size may be 2 feet by 1 foot 3 inches. When used for chops, ten or twelve can be cooked at one time ; or six chops can be cooked below the gas, and small pots boiled or stewed on top at the same time. They are very useful for making sauces, preserves, and things of like kind.

Steam Oven.—For potatoes, &c., the size may be 2 feet by 1 foot 6 inches by 2 feet 6 inches high, with two trays ; the steam pipe ¾ inch diameter. It will cook potatoes perfectly, the exact time can be regulated by the steam inlet cock. Other kinds of vegetables may also be cooked, except greens —these do not usually do so well.

Steam Jacketed Copper.—In some large establishments, where meat or soup is required to be cooked on a large scale, one of these may be found useful. The internal pan should be of copper, 10 lb. per super foot ; the exterior vessel or jacket may be cast iron ; the lid of copper hinged at the back, with chain and counterbalance for opening. The steam

pipe should be ¾ inch diameter. Pans for soup
making, &c., are sometimes hung upon trunnions.

The steam for the two apparatus may be obtained
from a small boiler set in connection with the
kitchen stove; or by one separately fired, placed
within the kitchen, or near it in another room.

It will be understood that any of these vessels
may be used or not, according to the requirements of
the place ; the steam oven and the jacketed copper
would only be required in very large places, where
a large number of servants have to be provided for.

Cooking by Gas is both economical and cleanly,
and effects a large saving of trouble. There are
manufacturers who have laid themselves out for
these special apparatus. The author believes the gas
stoves made by Fletcher and Co. to be the most
suitable and economical. Taking gas at an average
of 3s. per 1000 cubic feet, the saving between gas
and coal is considerable.

Both the gas and steam apparatus should be
placed under iron hoods, with free ventilation on top ;
the chimneys or pipes from the top of the hoods to
carry off the vapour should have a cowl or hood,
to prevent any down draught into the kitchen.

The Gas and Steam Cooking-room should have
the walls lined with glazed tiles, and the floor
should also be paved with tiles, and when this
kitchen is placed in an outer building, the top
should have a lantern light fitted with swing sashes
for top ventilation. No details can be given of the

general arrangement of this room, as it will differ according to the circumstances of the case.

VEGETABLE KITCHEN.

In large houses, a separate place should be provided for the preparation of all vegetables and salads. A washing sink made of slate, fitted up in much the same way as the kitchen sink, with a draining and cutting board, is about the only essential. In this room could be placed a small potato washing machine, and for large places a machine for washing greens and other vegetables.

BAKERY.

In large establishments in the country, the bread for all the household is generally made at home. Where a separate room can be had, it may be isolated from the house, and connected with it by a covered way open on one side.

The apparatus for bread making consists of " mixing-troughs," " kneading machines," and an " oven for baking."

Mixing-troughs may be made of wood or slate slabs ; hot and cold water should be laid on, and a plug and outlet pipe provided at the bottoms for washing out.

Kneading Machines. — The best kinds are Stevenson's ; they are very useful, and save much trouble. Bread can be entirely made in them, they are cleanly, and save much time as well as labour.

Ovens.—The best kind is Perkin's Hot Water Oven ; it is very economical. The oven is made with a double casing of iron ; the space, about 4 inches, is filled in with non-conducting composition to prevent escape of heat. The heating is done by *separate* wrought iron tubes partly filled with water, and closed at each end. The inclination of the tubes is towards the furnace, to keep the water at the fire end.

The tubes are fixed side by side in the oven. The rows are parallel ; one row is placed under the floor of the oven, and the other at the top inner surface. One end of each tube is projected into the fire. As the water becomes heated it expands and fills the tube with steam of equal temperature. The consumption of fuel is stated to be only one-tenth that used in ordinary fire ovens ; the fuel used costs about 2s. 10d. for each 100 lb. of bread. Each batch of bread can follow the last without any stoppage to re-heat the oven.

It costs but little to keep in repair. The baking temperature is from 420° to 460° F.; it is never heated above 500° F. The temperature is regulated by opening and closing the ash-pit door.

Ovens heated by Gas may also be employed ; they are worked as economically as those heated by coal or wood fires. These are a special kind of oven, and are fitted up successfully by Mr. W. J.

Booer, London. The cost of gas for each batch of bread in an average oven is about 9*d.* to 10*d.*

Ovens heated by Coal or Wood are not so good and give much more trouble than either Perkin's hot water system or the gas-heated system. When fire-heated ovens have to be adopted, it is better to put the construction in the hands of oven builders, who are experienced in this kind of work, so as to ensure the oven being properly set and economical in the consumption of fuel.

Flour should be contained in iron bins, and where there is sufficient height in the room they may be placed above the mixing troughs, out of head room, the bottom of the bins being made conical and fitted with a slide valve to discharge the quantity required. An automatic weighing machine may be fixed at the mouth of the hopper to weigh the flour if desired.

The rooms should have an open roof and a lantern at top, fitted with swing windows. The walls may be covered with glazed tiles, and the floor with red tiles set in cement. It is almost needless to say, the greatest cleanliness and thorough ventilation are essential, and that proper provision should be made for this in the building. The oven should, if possible, be placed in a separate room, to keep the bread-making room quite clear from dirt, dust, and smoke.

CHAPTER IX.

LAUNDRIES.

IN country houses and mansions it is usual to attach a laundry to wash the domestic and personal linen. The building should be entirely separate from the house. It may be constructed with two rooms, each of which should have an open roof, with lantern at the top fitted with swing sashes for ventilation. The first room—the wash-house—should have the floor either covered with asphalte or granolithic paving. Wood grids should be provided for the women to stand on at the washing troughs. The troughs for hot water may be made of deal covered with tinned copper sheets, the copper being well turned over the edges of the troughs and nailed with copper nails; a waste plug should be fitted at the bottom of each trough. Hot and cold water service should be provided. The troughs for cold water may be made of slate slabs 1 inch thick, fitted with waste outlet as before. Cold water service only should be laid on to these troughs. The waste water should be taken away by a wrought-iron pipe 2 inches diameter, and made to discharge over a trapped gully outside the house. At each outlet immediately under the trough a patent lead trap fitted with cleaning screw should be provided. The hot water can either be supplied from the house service, or a small boiler may be set in the laundry and used for this purpose only.

A Bradford's wringing machine should be provided. In large places a washing machine may be used ; these answer very well for sheets, cloths, and things of like kind. Those made by Bradford & Co. are the best, and do not injure the linen.

A centrifugal wringing machine may also be employed where large quantities of things have to be dried ; a machine with a basket 30 inches diameter is sufficient.

These machines usually require steam power to drive them. A small engine may be used, and supplied with steam from the boiler ; or in cases where only a hot water boiler and not a steam boiler is provided, a Davey's patent motor is a suitable engine to use. The washing machine can also be worked by this engine ; also the box and other mangles, as well as the steam calender, or other machines used.

A piece of light shafting may be run through the two rooms, fitted with bearings, hanging bracket, pulleys and belts. The machinery driven by power will of course only be needed in large establishments; the architect must determine if it should be used when he has learned the extent of the requirements of the place, and the amount of work to be done. In country mansions where a large staff of servants are employed there may be plenty of work that would justify such an outlay.

A Drying Stove should be provided ; this may be heated with steam pipes or hot air from the boiler furnace. The chambers may be two or three in number, according to the work to be done. The drying horses are run in and out of the chambers on wheels. At the top of each chamber a ventilating pipe should be provided; also an inlet of cold air at the bottom. The proper ventilation of these chambers is an important matter, and on it much of their success depends.

The Ironing Room should be provided with a gas stove for heating the irons, and in large places a steam calender may be provided for ironing the sheets and other large things; also a small ironing machine for collars, cuffs, and other light things.

The mangles may be of the roller kind made by Bradford & Co. Some people, however, prefer the old-fashioned box kind, especially for sheets, tablecloths, and other large things; the above firm also make an improved form of this machine. This room may either be heated by an open fire, or by hot water or steam pipes, the former is usually preferred. The suggestions here given are general; the extent to which they may be applied must rest with the judgment of the architect. The requirements of a good modern sanitary laundry have been described, and, in the author's opinion, this is a very essential department in every well-regulated country residence or mansion, as well as in boarding houses and large residential mansions.

Any one wanting further information of apparatus and machines for large public institutions is referred to the author's book upon " Engineering and other Appliances for Public Institutions ": E. and F. N. Spon, London.

CHAPTER X.

GREENHOUSES AND HOTHOUSES.

THE means of heating these, and their ventilation, will be described; the outlines of the general construction will also be treated, and a few suggestions thrown out as to some details to which attention should be given to ensure a successful result in working. The houses and apparatus described are only for private owners, and not for market gardeners.

The Heating Apparatus. — The arrangements will much depend upon the relative position and number of the houses to be heated; when several are placed close to each other, one boiler can be used to work all of them. The best kind of boiler is the wrought-iron saddle type, or the Trentham or Cornish boiler type; for *small* houses a vertical boiler fired from the top may be used. The size of the boiler depends upon the length of pipes required, the makers usually state what power the boilers will give; it is better, however, to have one of ample capacity, so as not to force it, as this is false economy. The boiler must be fixed in brickwork, and below the floor of the houses, to give space for the pipes to rise from the *top* of the boiler and pass under the floor without dipping down.

G

The pipes used are 4 inches diameter, of the socket kind ; the joints should be made with gasket and red and white lead. The " flow " or supply pipe is taken from the *top* of the boiler, and the " return " is taken back to the *bottom* of it. The pipes must be fixed on the rise *from* the top of the boiler and *fall* back to the bottom of it, they must not be dipped in any part ; at the bends, air cocks must be provided to keep them clear of air ; these must be opened occasionally, especially at the first starting each season of the apparatus. Throttle valves must be provided in the pipes to shut off or to regulate the quantity of water as required. The pipes have also to be provided with expansion joints to allow for expansion and contraction, otherwise the joints will be broken. Roller carriages set in frames are provided for the pipes to rest and *move* on—they must *not* be built anywhere in the walls.

A relief or expansion pipe must be fitted to the " flow " pipe in each house, and carried to the outside of the roof.

The cubical contents of each house must be taken, and the following quantity of pipe allowed, according to the temperature required : — For greenhouses, 35 feet run of 4 inch pipes per 1000 cubic feet for a temperature of 55°. Hot houses and graperies per 1000 cubic feet, say 45 feet run of 4 inch pipes for a temperature of 70°. For forcing houses and cucumber houses per 1000 cubic feet, say 55 feet run of 4 inch pipes for a temperature of 80°. The time for heating up in this latter case is about three-quarters of an hour, the coke used is about 2 to 2½ bushels in 10 hours.

Note.—If 3 inch diameter pipes be used, increase the above quantities one-third.

It must be borne in mind that a most perfect *circulation* of water must be maintained in the pipes, they must *not* be passed *under doorways and then made to rise again.*

When the length of the pipe is determined for each house, it must then be decided how many lines of coils are necessary. The best position for the lines of pipes is next the walls under the benches ; but in cases where they have to be passed from house to house, it may sometimes be more convenient to put them in trenches, with perforated gratings to cover them.

For Conservatories, where not much heat is required, coil cases or radiators may be used ; as the pipes in this case are not continuous, a rather larger allowance as to the length must be made. In most cases 2 inch diameter pipes will be the most suitable, as the heat can be got up more quickly ; the quantity of 2 inch pipe required is double that of 4 inch.

For Forcing Houses, the hot water pipes are usually laid under the mould of the bed. A good allowance of pipe must be made, as at times great heat is wanted ; it is therefore essential to have some reserve, as it can easily be regulated to suit requirements.

Size of the Pipes.—It should be stated as a rule 4 inch diameter are the best to use, because they cool less quickly than the smaller sizes; the rate of cooling is in the following proportion :—

Diameter of pipes	..	2 inch	3 inch	4 inch.
Rates of cooling	..	2	1·3	1

Coil Cases.—It is often a convenience, especially in conservatories, to use 3 inch diameter pipes; in this case increase the quantity one-third more than for 4 inch diameter, and if 2 inch diameter are used, double the quantity required for 4 inch diameter. It will be understood that the temperature can be raised more *quickly* with the small sized pipes, but they *cool* much quicker.

Boilers must be set in brickwork by men accustomed to this kind of work. The area of the fire grate must be one-sixth the heating surface of the boiler. The quantity of air required through the grate bars is 10 to 11 lbs. per 1 lb. of coal. The quantity of water heated by the boiler for each square foot of heating surface is 11 gallons per hour from a temperature of 52° to 212°. 1 lb. of coal will heat 39 to 40 lbs. of water from 32° to 212°. One foot run of 4 inch diameter pipe will heat 222 cubic feet of air 1° per minute when the difference of the temperature between the pipes and the air is 125°. A safety valve should be provided at the boiler in case the relief pipes do not act.

Pipes with patent joints made with india-rubber are preferred by some people; they make a good job, but are not any better than the ordinary socket joints. The quality of the pipes requires much attention; they should only be obtained of the first-class makers; cheapness in their purchase will surely lead to trouble in after working.

Special Valves are made to shut off the water from any part; the throttle valves before described are only used to control the quantity running through any particular line of pipes.

Thermometers should be placed in different parts of the houses to enable the gardener to regulate the heat as required. It is also advisable to have a special one sunk in the "flow" pipes at the entry of each house. This shows the temperature of the water inside the pipes, and is a good test to see that the apparatus is working properly.

The Ventilation of the houses requires the most careful attention, both for the inlet as well as the outlet of air. The sashes at the sides and at the roof should be opened and shut by automatic adjustable gear made of iron. In addition to this, it is advisable to fix a few "hit and miss" ventilators, to let in small quantities of fresh air; the number of these will depend upon the size of the house.

Provision should be made in each house to obtain warm water for watering the plants. There are several simple means of accomplishing this.

Cold-water Pipes and Cocks, also indiarubber

hoses, should be provided for washing down, &c. In places where good pressure can be obtained from the main pipes, the hose may be used for syringing the plants.

Greenhouses and Hothouses are now often constructed entirely of iron instead of wood; they are of necessity more costly, but last a longer time. The glass should be secured by red lead; ordinary putty does not last any time. Patent sash-bars of wrought iron are now made covered with sheet lead; the glass rests on a rounded edge, and when in its place the lead is turned down over it at each side. These bars are made by the "British Patent Glazing Company." Rendle's patent system of glazing may also be applied to greenhouses, &c. with much advantage.

Houses constructed with a double-pitch roof are, as a rule, preferred to lean-to houses; the latter are, however, more economical; and when plants require training on the walls, they are a necessity.

Graperies are generally preferred constructed with a lean-to roof. Wrought-iron rods should be provided and fixed directly under the lights, to which the vine is hung. A hot-bed can be formed about 2 feet 6 inches to 3 feet high above the floor, in which some plants may be grown.

Most gardeners prefer to have a bed 8 to 9 feet wide left outside the house for the roots of the vines. There seems now to be some difference of opinion as to the expediency of this.

Cucumber-Houses.—The floor should be sunk below the ground level, and the side lights brought down nearly to the level of the ground. Lean-to houses appear to be the most suitable ; they should be fitted up with rods attached to the lights in the same manner as for graperies. The hot-bed is made about 2 feet 6 inches to 3 feet high above the floor, the hot-water pipes being fixed under the mould. The bine of the plants is trained or hung under the iron rods, close to the lights.

Melon-Houses and Orchard-Houses are constructed in much the same manner. It is not proposed to enter into any detail as to these, except to note that ample power of heating should be provided. In the case of orchard-houses, the height will mainly depend upon the plants to be grown in them.

The Floors of all the houses above described may be laid with Stuart's patent granolithic concrete. The passages or pathways are laid with coloured tiles, where the extra expense is no object. It is a matter of some importance to have the houses so constructed that they can be washed down and kept perfectly clean. This, and efficient ventilation, as well as perfect heating, are the chief essentials for making a good and successful house.

As before stated, it is not intended to enter into close detail as to the construction of houses, except to give a few hints as to general principles that should guide the designer.

The author advises the architect to get particulars of what is wanted from those in authority. If he carries out their general ideas as far as he can, they will be responsible for any defect in the working of the place. These suggestions are made because gardeners as a class are most difficult to please, and are full of crotchets and fads as to the best kind of houses and apparatus.

It is hoped that the suggestions given here will be found useful, and act as a guide to professional men who may not have had experience in this branch of practice. At any rate, it may save them from having to place themselves absolutely in the hands of greenhouse builders, some of whom are not always to be relied on, especially when there is any competition to get the work. In this case, too often, good design and efficient working are sacrificed to cheapness and bad work.

As regards the heating apparatus, it should always be placed in the hands of contractors who have had special experience in this class of work. The author has employed the old-established firm of Bailey, Pegg, & Co., on many occasions, and has always been well satisfied with the materials and the workmanship. Added to this, the work has always turned out perfectly successful, which is the most important consideration, as far as the architect is concerned, and saves all further dissatisfaction on the part of the client.

INDEX.

H

LONDON: PRINTED BY WILLIAM CLOWES AND SONS, LIMITED,

BOOKS RELATING

TO

APPLIED SCIENCE,

PUBLISHED BY

E. & F. N. SPON,

LONDON: 125, STRAND.

NEW YORK: 12, CORTLANDT STREET.

———◆———

A Pocket-Book for Chemists, Chemical Manufacturers,
Metallurgists, Dyers, Distillers, Brewers, Sugar Refiners, Photographers,
Students, etc., etc. By THOMAS BAYLEY, Assoc. R.C. Sc. Ireland, Ana-
lytical and Consulting Chemist and Assayer. Fourth edition, with
additions, 437 pp., royal 32mo, roan, gilt edges, 5s.

SYNOPSIS OF CONTENTS :

Atomic Weights and Factors—Useful Data—Chemical Calculations—Rules for Indirect
Analysis—Weights and Measures—Thermometers and Barometers—Chemical Physics—
Boiling Points, etc.—Solubility of Substances—Methods of Obtaining Specific Gravity—Con-
version of Hydrometers—Strength of Solutions by Specific Gravity—Analysis—Gas Analysis—
Water Analysis—Qualitative Analysis and Reactions—Volumetric Analysis—Manipulation—
Mineralogy—Assaying—Alcohol—Beer—Sugar—Miscellaneous Technological matter
relating to Potash, Soda, Sulphuric Acid, Chlorine, Tar Products, Petroleum, Milk, Tallow,
Photography, Prices, Wages, Appendix, etc., etc.

The Mechanician : A Treatise on the Construction
and Manipulation of Tools, for the use and instruction of Young Engineers
and Scientific Amateurs, comprising the Arts of Blacksmithing and Forging ; the Construction and Manufacture of Hand Tools, and the various
Methods of Using and Grinding them ; the Construction of Machine Tools,
and how to work them ; Machine Fitting and Erection ; description of
Hand and Machine Processes ; Turning and Screw Cutting ; principles of
Constructing and details of Making and Erecting Steam Engines, and the
various details of setting out work, etc., etc. By CAMERON KNIGHT,
Engineer. *Containing* 1147 *illustrations,* and 397 pages of letter-press,
Fourth edition, 4to, cloth, 18s.

B

Just Published, in Demy 8vo, cloth, containing 975 *pages and* 250 *Illustrations, price* 7s. 6d.

SPONS' HOUSEHOLD MANUAL:

A Treasury of Domestic Receipts and Guide for Home Management.

~~~~~~~~~~~~~~~

PRINCIPAL CONTENTS.

Hints for selecting a good House, pointing out the essential requirements for a good house as to the Site, Soil, Trees, Aspect, Construction, and General Arrangement; with instructions for Reducing Echoes, Waterproofing Damp Walls, Curing Damp Cellars.

Sanitation.—What should constitute a good Sanitary Arrangement; Examples (with illustrations) of Well- and Ill-drained Houses; How to Test Drains; Ventilating Pipes, etc.

Water Supply.—Care of Cisterns; Sources of Supply; Pipes; Pumps; Purification and Filtration of Water.

Ventilation and Warming.—Methods of Ventilating without causing cold draughts, by various means; Principles of Warming; Health Questions; Combustion; Open Grates; Open Stoves; Fuel Economisers; Varieties of Grates; Close-Fire Stoves; Hot-air Furnaces; Gas Heating; Oil Stoves; Steam Heating; Chemical Heaters; Management of Flues; and Cure of Smoky Chimneys.

Lighting.—The best methods of Lighting; Candles, Oil Lamps, Gas, Incandescent Gas, Electric Light; How to test Gas Pipes; Management of Gas.

Furniture and Decoration.—Hints on the Selection of Furniture; on the most approved methods of Modern Decoration; on the best methods of arranging Bells and Calls; How to Construct an Electric Bell.

Thieves and Fire.—Precautions against Thieves and Fire; Methods of Detection; Domestic Fire Escapes; Fireproofing Clothes, etc.

The Larder.—Keeping Food fresh for a limited time; Storing Food without change, such as Fruits, Vegetables, Eggs, Honey, etc.

Curing Foods for lengthened Preservation, as Smoking, Salting, Canning, Potting, Pickling, Bottling Fruits, etc.; Jams, Jellies, Marmalade, etc.

The Dairy.—The Building and Fitting of Dairies in the most approved modern style; Butter-making; Cheesemaking and Curing.

The Cellar.—Building and Fitting; Cleaning Casks and Bottles; Corks and Corking; Aërated Drinks; Syrups for Drinks; Beers; Bitters; Cordials and Liqueurs; Wines; Miscellaneous Drinks.

The Pantry.—Bread-making; Ovens and Pyrometers; Yeast; German Yeast; Biscuits; Cakes; Fancy Breads; Buns.

The Kitchen.—On Fitting Kitchens; a description of the best Cooking Ranges, close and open; the Management and Care of Hot Plates, Baking Ovens, Dampers, Flues, and Chimneys; Cooking by Gas; Cooking by Oil; the Arts of Roasting, Grilling, Boiling, Stewing, Braising, Frying.

Receipts for Dishes.—Soups, Fish, Meat, Game, Poultry, Vegetables, Salads, Puddings, Pastry, Confectionery, Ices, etc., etc.; Foreign Dishes.

The Housewife's Room.—Testing Air, Water, and Foods; Cleaning and Renovating; Destroying Vermin.

Housekeeping, Marketing.

The Dining-Room.—Dietetics; Laying and Waiting at Table; Carving; Dinners, Breakfasts, Luncheons, Teas, Suppers, etc.

The Drawing-Room.—Etiquette; Dancing; Amateur Theatricals; Tricks and Illusions; Games (indoor).

The Bedroom and Dressing-Room; Sleep; the Toilet; Dress; Buying Clothes; Outfits; Fancy Dress.

The Nursery.—The Room; Clothing; Washing; Exercise; Sleep; Feeding; Teething; Illness; Home Training.

The Sick-Room.—The Room; the Nurse; the Bed; Sick Room Accessories; Feeding Patients; Invalid Dishes and Drinks; Administering Physic; Domestic Remedies; Accidents and Emergencies; Bandaging; Burns; Carrying Injured Persons; Wounds; Drowning; Fits; Frost-bites; Poisons and Antidotes; Sunstroke; Common Complaints; Disinfection, etc.

The Bath-Room.—Bathing in General; Management of Hot-Water System.

The Laundry.—Small Domestic Washing Machines, and methods of getting up linen; Fitting up and Working a Steam Laundry.

The School-Room.—The Room and its Fittings; Teaching, etc.

The Playground.—Air and Exercise; Training; Outdoor Games and Sports.

The Workroom.—Darning, Patching, and Mending Garments.

The Library.—Care of Books.

The Garden.—Calendar of Operations for Lawn, Flower Garden, and Kitchen Garden.

The Farmyard.—Management of the Horse, Cow, Pig, Poultry, Bees, etc., etc.

Small Motors.—A description of the various small Engines useful for domestic purposes, from 1 man to 1 horse power, worked by various methods, such as Electric Engines, Gas Engines, Petroleum Engines, Steam Engines, Condensing Engines, Water Power, Wind Power, and the various methods of working and managing them.

Household Law.—The Law relating to Landlords and Tenants, Lodgers, Servants, Parochial Authorities, Juries, Insurance, Nuisance, etc.

On Designing Belt Gearing. By E. J. COWLING
WELCH, Mem. Inst. Mech. Engineers, Author of 'Designing Valve Gearing.' Fcap. 8vo, sewed, 6d.

A Handbook of Formulæ, Tables, and Memoranda,
for Architectural Surveyors and others engaged in Building. By J. T. HURST, C.E. Fourteenth edition, royal 32mo, roan, 5s.

"It is no disparagement to the many excellent publications we refer to, to say that in our opinion this little pocket-book of Hurst's is the very best of them all, without any exception. It would be useless to attempt a recapitulation of the contents, for it appears to contain almost *everything* that anyone connected with building could require, and, best of all, made up in a compact form for carrying in the pocket, measuring only 5 in. by 3 in., and about ¼ in. thick, in a limp cover. We congratulate the author on the success of his laborious and practically compiled little book, which has received unqualified and deserved praise from every professional person to whom we have shown it."—*The Dublin Builder.*

Tabulated Weights of Angle, Tee, Bulb, Round,
Square, and Flat Iron and Steel, and other information for the use of Naval Architects and Shipbuilders. By C. H. JORDAN, M.I.N.A. Fourth edition, 32mo, cloth, 2s. 6d.

A Complete Set of Contract Documents for a Country
Lodge, comprising Drawings, Specifications, Dimensions (for quantities), Abstracts, Bill of Quantities, Form of Tender and Contract, with Notes by J. LEANING, printed in facsimile of the original documents, on single sheets fcap., in paper case, 10s.

A Practical Treatise on Heat, as applied to the
Useful Arts; for the Use of Engineers, Architects, &c. By THOMAS BOX. With 14 plates. Third edition, crown 8vo, cloth, 12s. 6d.

A Descriptive Treatise on Mathematical Drawing
Instruments: their construction, uses, qualities, selection, preservation, and suggestions for improvements, with hints upon Drawing and Colouring. By W. F. STANLEY, M.R.I. Fifth edition, *with numerous illustrations*, crown 8vo, cloth, 5s.

Quantity Surveying. By J. LEANING. With 42 illustrations. Second edition, revised, crown 8vo, cloth, 9s.

CONTENTS :

A complete Explanation of the London Practice.
General Instructions.
Order of Taking Off.
Modes of Measurement of the various Trades.
Use and Waste.
Ventilation and Warming,
Credits, with various Examples of Treatment.
Abbreviations.
Squaring the Dimensions
Abstracting, with Examples in illustration of each Trade.
Billing.
Examples of Preambles to each Trade.
Form for a Bill of Quantities.
   Do.   Bill of Credits.
   Do.   Bill for Alternative Estimate.
Restorations and Repairs, and Form of Bill.
Variations before Acceptance of Tender.
Errors in a Builder's Estimate.

Schedule of Prices.
Form of Schedule of Prices.
Analysis of Schedule of Prices.
Adjustment of Accounts.
Form of a Bill of Variations.
Remarks on Specifications.
Prices and Valuation of Work, with Examples and Remarks upon each Trade.
The Law as it affects Quantity Surveyors, with Law Reports.
Taking Off after the Old Method.
Northern Practice.
The General Statement of the Methods recommended by the Manchester Society of Architects for taking Quantities.
Examples of Collections.
Examples of " Taking Off" in each Trade.
Remarks on the Past and Present Methods of Estimating.

Spons' Architects' and Builders' Price Book, with useful Memoranda. Edited by W. YOUNG, Architect. Crown 8vo, cloth, red edges, 3s. 6d. *Published annually.* Sixteenth edition. *Now ready.*

Long-Span Railway Bridges, comprising Investigations of the Comparative Theoretical and Practical Advantages of the various adopted or proposed Type Systems of Construction, with numerous Formulæ and Tables giving the weight of Iron or Steel required in Bridges from 300 feet to the limiting Spans ; to which are added similar Investigations and Tables relating to Short-span Railway Bridges. Second and revised edition. By B. BAKER, Assoc. Inst. C.E. *Plates,* crown 8vo, cloth, 5s.

Elementary Theory and Calculation of Iron Bridges and Roofs. By AUGUST RITTER, Ph.D., Professor at the Polytechnic School at Aix-la-Chapelle. Translated from the third German edition, by H. R. SANKEY, Capt. R.E. With 500 *illustrations,* 8vo, cloth, 15s.

The Elementary Principles of Carpentry. By THOMAS TREDGOLD. Revised from the original edition, and partly re-written, by JOHN THOMAS HURST. Contained in 517 pages of letter-press, and *illustrated with* 48 *plates and* 150 *wood engravings.* Sixth edition, reprinted from the third, crown 8vo, cloth, 12s. 6d.

Section I. On the Equality and Distribution of Forces — Section II. Resistance of Timber — Section III. Construction of Floors — Section IV. Construction of Roofs — Section V. Construction of Domes and Cupolas — Section VI. Construction of Partitions — Section VII. Scaffolds, Staging, and Gantries — Section VIII. Construction of Centres for Bridges — Section IX. Coffer-dams, Shoring, and Strutting — Section X. Wooden Bridges and Viaducts — Section XI. Joints, Straps, and other Fastenings — Section XII. Timber.

The Builder's Clerk : a Guide to the Management of a Builder's Business. By THOMAS BALES. Fcap. 8vo, cloth, 1s. 6d.

Our Factories, Workshops, and Warehouses: their
Sanitary and Fire-Resisting Arrangements. By B. H. THWAITE, Assoc.
Mem. Inst. C.E. *With 183 wood engravings,* crown 8vo, cloth, 9*s.*

Hot Water Supply: A Practical Treatise upon the
Fitting of Circulating Apparatus in connection with Kitchen Range and
other Boilers, to supply Hot Water for Domestic and General Purposes.
With a Chapter upon Estimating. *Fully illustrated,* crown 8vo, cloth, 3*s.*

Hot Water Apparatus: An Elementary Guide for
the Fitting and Fixing of Boilers and Apparatus for the Circulation of
Hot Water for Heating and for Domestic Supply, and containing a
Chapter upon Boilers and Fittings for Steam Cooking. 32 *illustrations,*
fcap. 8vo, cloth, 1*s.* 6*d.*

The Use and Misuse, and the Proper and Improper
Fixing of a Cooking Range. Illustrated, fcap. 8vo, sewed, 6*d.*

Iron Roofs: Examples of Design, Description. *Illus-*
trated with 64 *Working Drawings of Executed Roofs.* By ARTHUR T.
WALMISLEY, Assoc. Mem. Inst. C.E. Second edition, revised, imp. 4to,
half-morocco, 3*l.* 3*s.*

A History of Electric Telegraphy, to the Year 1837.
Chiefly compiled from Original Sources, and hitherto Unpublished Documents, by J. J. FAHIE, Mem. Soc. of Tel. Engineers, and of the Inter-
national Society of Electricians, Paris. Crown 8vo, cloth, 9*s.*

Spons' Information for Colonial Engineers. Edited
by J. T. HURST. Demy 8vo, sewed.

No. 1, Ceylon. By ABRAHAM DEANE, C.E. 2*s.* 6*d.*

CONTENTS :

Introductory Remarks — Natural Productions — Architecture and Engineering — Topography, Trade, and Natural History—Principal Stations—Weights and Measures, etc., etc.

No. 2. Southern Africa, including the Cape Colony, Natal, and the
Dutch Republics. By HENRY HALL, F.R.G.S., F.R.C.I. With
Map. 3*s.* 6*d.*

CONTENTS :

General Description of South Africa—Physical Geography with reference to Engineering
Operations—Notes on Labour and Material in Cape Colony—Geological Notes on Rock
Formation in South Africa—Engineering Instruments for Use in South Africa—Principal
Public Works in Cape Colony: Railways, Mountain Roads and Passes, Harbour Works,
Bridges, Gas Works, Irrigation and Water Supply, Lighthouses, Drainage and Sanitary
Engineering, Public Buildings, Mines—Table of Woods in South Africa—Animals used for
Draught Purposes—Statistical Notes—Table of Distances—Rates of Carriage, etc.

No. 3. India. By F. C. DANVERS, Assoc. Inst. C.E. With Map. 4*s.* 6*d.*

CONTENTS :

Physical Geography of India—Building Materials—Roads—Railways—Bridges—Irrigation — River Works — Harbours — Lighthouse Buildings — Native Labour — The Principal
Trees of India—Money—Weights and Measures—Glossary of Indian Terms, etc.

A Practical Treatise on Coal Mining. By GEORGE
G. ANDRÉ, F.G.S., Assoc. Inst. C.E., Member of the Society of Engineers.
With 82 lithographic plates. 2 vols., royal 4to, cloth, 3*l.* 12*s.*

A Practical Treatise on Casting and Founding,
including descriptions of the modern machinery employed in the art. By
N. E. SPRETSON, Engineer. Third edition, with 82 *plates* drawn to
scale, 412 pp., demy 8vo, cloth, 18*s.*

The Depreciation of Factories and their Valuation.
By EWING MATHESON, M. Inst. C.E. 8vo, cloth, 6*s.*

A Handbook of Electrical Testing. By H. R. KEMPE,
M.S.T.E. Fourth edition, revised and enlarged, crown 8vo, cloth, 16*s.*

Gas Works : their Arrangement, Construction, Plant,
and Machinery. By F. COLYER, M. Inst. C.E. *With 31 folding plates,*
8vo, cloth, 24*s.*

The Clerk of Works : a Vade-Mecum for all engaged
in the Superintendence of Building Operations. By G. G. HOSKINS,
F.R.I.B.A. Third edition, fcap. 8vo, cloth, 1*s.* 6*d.*

American Foundry Practice : Treating of Loam,
Dry Sand, and Green Sand Moulding, and containing a Practical Treatise
upon the Management of Cupolas, and the Melting of Iron. By T. D.
WEST, Practical Iron Moulder and Foundry Foreman. Second edition,
with numerous illustrations, crown 8vo, cloth, 10*s.* 6*d.*

The Maintenance of Macadamised Roads. By T.
CODRINGTON, M.I.C.E, F.G.S., General Superintendent of County Roads
for South Wales. 8vo, cloth, 6*s.*

Hydraulic Steam and Hand Power Lifting and
Pressing Machinery. By FREDERICK COLYER, M. Inst. C.E., M. Inst. M.E.
With 73 plates, 8vo, cloth, 18*s.*

Pumps and Pumping Machinery. By F. COLYER,
M.I.C.E., M.I.M.E. *With 23 folding plates,* 8vo, cloth, 12*s.* 6*d.*

Pumps and Pumping Machinery. By F. COLYER.
Second Part. *With 11 large plates,* 8vo, cloth, 12*s.* 6*d.*

A Treatise on the Origin, Progress, Prevention, and
Cure of Dry Rot in Timber; with Remarks on the Means of Preserving
Wood from Destruction by Sea-Worms, Beetles, Ants, etc. By THOMAS
ALLEN BRITTON, late Surveyor to the Metropolitan Board of Works,
etc., etc. *With 10 plates,* crown 8vo, cloth, 7*s.* 6*d.*

The Municipal and Sanitary Engineer's Handbook.

By H. PERCY BOULNOIS, Mem. Inst. C.E., Borough Engineer, Portsmouth. *With numerous illustrations*, demy 8vo, cloth, 12s. 6d.

CONTENTS:

The Appointment and Duties of the Town Surveyor—Traffic—Macadamised Roadways—Steam Rolling—Road Metal and Breaking—Pitched Pavements—Asphalte—Wood Pavements—Footpaths—Kerbs and Gutters—Street Naming and Numbering—Street Lighting—Sewerage—Ventilation of Sewers—Disposal of Sewage—House Drainage—Disinfection—Gas and Water Companies, etc., Breaking up Streets—Improvement of Private Streets—Borrowing Powers—Artizans' and Labourers' Dwellings—Public Conveniences—Scavenging, including Street Cleansing—Watering and the Removing of Snow—Planting Street Trees—Deposit of Plans—Dangerous Buildings—Hoardings—Obstructions—Improving Street Lines—Cellar Openings—Public Pleasure Grounds—Cemeteries—Mortuaries—Cattle and Ordinary Markets—Public Slaughter-houses, etc.—Giving numerous Forms of Notices, Specifications, and General Information upon these and other subjects of great importance to Municipal Engineers and others engaged in Sanitary Work.

Metrical Tables. By G. L. MOLESWORTH, M.I.C.E.

32mo, cloth, 1s. 6d.

CONTENTS.

General—Linear Measures—Square Measures—Cubic Measures—Measures of Capacity—Weights—Combinations—Thermometers.

Elements of Construction for Electro-Magnets. By

Count TH. DU MONCEL, Mem. de l'Institut de France. Translated from the French by C. J. WHARTON. Crown 8vo, cloth, 4s. 6d.

Practical Electrical Units Popularly Explained, with

numerous illustrations and Remarks. By JAMES SWINBURNE, late of J. W. Swan and Co., Paris, late of Brush-Swan Electric Light Company, U.S.A. 18mo, cloth, 1s. 6d.

A Treatise on the Use of Belting for the Transmission of Power. By J. H. COOPER. Second edition, *illustrated*, 8vo, cloth, 15s.

A Pocket-Book of Useful Formulæ and Memoranda

for Civil and Mechanical Engineers. By GUILFORD L. MOLESWORTH, Mem. Inst. C.E., Consulting Engineer to the Government of India for State Railways. *With numerous illustrations*, 744 pp. Twenty-second edition, revised and enlarged, 32mo, roan, 6s.

SYNOPSIS OF CONTENTS:

Surveying, Levelling, etc.—Strength and Weight of Materials—Earthwork, Brickwork, Masonry, Arches, etc.—Struts, Columns, Beams, and Trusses—Flooring, Roofing, and Roof Trusses—Girders, Bridges, etc.—Railways and Roads—Hydraulic Formulæ—Canals, Sewers, Waterworks, Docks—Irrigation and Breakwaters—Gas, Ventilation, and Warming—Heat, Light, Colour, and Sound—Gravity: Centres, Forces, and Powers—Millwork, Teeth of Wheels, Shafting, etc.—Workshop Recipes—Sundry Machinery—Animal Power—Steam and the Steam Engine—Water-power, Water-wheels, Turbines, etc.—Wind and Windmills—Steam Navigation, Ship Building, Tonnage, etc.—Gunnery, Projectiles, etc.—Weights, Measures, and Money—Trigonometry, Conic Sections, and Curves—Telegraphy—Mensuration—Tables of Areas and Circumference, and Arcs of Circles—Logarithms, Square and Cube Roots, Powers—Reciprocals, etc.—Useful Numbers—Differential and Integral Calculus—Algebraic Signs—Telegraphic Construction and Formulæ.

Hints on Architectural Draughtsmanship. By G. W. TUXFORD HALLATT. Fcap. 8vo, cloth, 1s. 6d.

Spons' Tables and Memoranda for Engineers; selected and arranged by J. T. HURST, C.E., Author of 'Architectural Surveyors' Handbook,' 'Hurst's Tredgold's Carpentry,' etc. Ninth edition, 64mo, roan, gilt edges, 1s.; or in cloth case, 1s. 6d.

This work is printed in a pearl type, and is so small, measuring only 2½ in. by 1¾ in. by ¾ in. thick, that it may be easily carried in the waistcoat pocket.

"It is certainly an extremely rare thing for a reviewer to be called upon to notice a volume measuring but 2½ in. by 1¾ in., yet these dimensions faithfully represent the size of the handy little book before us. The volume—which contains 118 printed pages, besides a few blank pages for memoranda—is, in fact, a true pocket-book, adapted for being carried in the waistcoat pocket, and containing a far greater amount and variety of information than most people would imagine could be compressed into so small a space. . . . . The little volume has been compiled with considerable care and judgment, and we can cordially recommend it to our readers as a useful little pocket companion."—*Engineering.*

A Practical Treatise on Natural and Artificial Concrete, its Varieties and Constructive Adaptations. By HENRY REID, Author of the 'Science and Art of the Manufacture of Portland Cement.' New Edition, *with 59 woodcuts and 5 plates,* 8vo, cloth, 15s.

Notes on Concrete and Works in Concrete; especially written to assist those engaged upon Public Works. By JOHN NEWMAN, Assoc. Mem. Inst. C.E., crown 8vo, cloth, 4s. 6d.

Electricity as a Motive Power. By Count TH. DU MONCEL, Membre de l'Institut de France, and FRANK GERALDY, Ingénieur des Ponts et Chaussées. Translated and Edited, with Additions, by C. J. WHARTON, Assoc. Soc. Tel. Eng. and Elec. *With 113 engravings and diagrams,* crown 8vo, cloth, 7s. 6d.

Treatise on Valve-Gears, with special consideration of the Link-Motions of Locomotive Engines. By Dr. GUSTAV ZEUNER, Professor of Applied Mechanics at the Confederated Polytechnikum of Zurich. Translated from the Fourth German Edition, by Professor J. F. KLEIN, Lehigh University, Bethlehem, Pa. *Illustrated,* 8vo, cloth, 12s. 6d.

The French-Polisher's Manual. By a French-Polisher; containing Timber Staining, Washing, Matching, Improving, Painting, Imitations, Directions for Staining, Sizing, Embodying, Smoothing, Spirit Varnishing, French-Polishing, Directions for Re-polishing. Third edition, royal 32mo, sewed, 6d.

Hops, their Cultivation, Commerce, and Uses in various Countries. By P. L. SIMMONDS. Crown 8vo, cloth, 4s. 6d.

The Principles of Graphic Statics. By GEORGE SYDENHAM CLARKE, Capt. Royal Engineers. *With 112 illustrations.* 4to, cloth, 12s. 6d.

Dynamo-Electric Machinery : A Manual for Students of Electro-technics. By SILVANUS P. THOMPSON, B.A., D.Sc., Professor of Experimental Physics in University College, Bristol, etc., etc. Third edition, *illustrated,* 8vo, cloth, 16s.

Practical Geometry,• Perspective, and Engineering Drawing; a Course of Descriptive Geometry adapted to the Requirements of the Engineering Draughtsman, including the determination of cast shadows and Isometric Projection, each chapter being followed by numerous examples ; to which are added rules for Shading, Shade-lining, etc., together with practical instructions as to the Lining, Colouring, Printing, and general treatment of Engineering Drawings, with a chapter on drawing Instruments. By GEORGE S. CLARKE, Capt. R.E. Second edition, *with 21 plates.* 2 vols., cloth, 10s. 6d.

The Elements of Graphic Statics. By Professor KARL VON OTT, translated from the German by G. S. CLARKE, Capt. R.E., Instructor in Mechanical Drawing, Royal Indian Engineering College. *With 93 illustrations,* crown 8vo, cloth, 5s.

A Practical Treatise on the Manufacture and Distribution of Coal Gas. By WILLIAM RICHARDS. Demy 4to, with *numerous wood engravings and 29 plates,* cloth, 28s.

SYNOPSIS OF CONTENTS :

Introduction—History of Gas Lighting—Chemistry of Gas Manufacture, by Lewis Thompson, Esq., M.R.C.S.—Coal, with Analyses, by J. Paterson, Lewis Thompson, and G. R. Hislop, Esqrs.—Retorts, Iron and Clay—Retort Setting—Hydraulic Main—Condensers—Exhausters—Washers and Scrubbers—Purifiers—Purification—History of Gas Holder—Tanks, Brick and Stone, Composite, Concrete, Cast-iron, Compound Annular Wrought-iron—Specifications—Gas Holders—Station Meter—Governor—Distribution—Mains—Gas Mathematics, or Formulæ for the Distribution of Gas, by Lewis Thompson, Esq.—Services—Consumers' Meters—Regulators—Burners—Fittings—Photometer—Carburization of Gas—Air Gas and Water Gas—Composition of Coal Gas, by Lewis Thompson, Esq.—Analyses of Gas—Influence of Atmospheric Pressure and Temperature on Gas—Residual Products—Appendix—Description of Retort Settings, Buildings, etc., etc.

The New Formula for Mean Velocity of Discharge of Rivers and Canals. By W. R. KUTTER. Translated from articles in the 'Cultur-Ingénieur,' by LOWIS D'A. JACKSON, Assoc. Inst. C.E. 8vo, cloth, 12s. 6d.

The Practical Millwright and Engineer's Ready Reckoner; or Tables for finding the diameter and power of cog-wheels, diameter, weight, and power of shafts, diameter and strength of bolts, etc. By THOMAS DIXON. Fourth edition, 12mo, cloth, 3s.

Tin : Describing the Chief Methods of Mining, Dressing and Smelting it abroad ; with Notes upon Arsenic, Bismuth and Wolfram. By ARTHUR G. CHARLETON, Mem. American Inst. of Mining Engineers. *With plates,* 8vo, cloth, 12s. 6d.

Perspective, Explained and Illustrated. By G. S.
CLARKE, Capt. R.E. *With illustrations,* 8vo, cloth, 3s. 6d.

Practical Hydraulics ; a Series of Rules and Tables
for the use of Engineers, etc., etc. By THOMAS BOX. Fifth edition,
numerous plates, post 8vo, cloth, 5s.

The Essential Elements of Practical Mechanics ;
based on the Principle of Work, designed for Engineering Students. By
OLIVER BYRNE, formerly Professor of Mathematics, College for Civil
Engineers. Third edition, *with* 148 *wood engravings,* post 8vo, cloth,
7s. 6d.

CONTENTS :

Chap. 1. How Work is Measured by a Unit, both with and without reference to a Unit
of Time—Chap. 2. The Work of Living Agents, the Influence of Friction, and introduces
one of the most beautiful Laws of Motion—Chap. 3. The principles expounded in the first and
second chapters are applied to the Motion of Bodies—Chap. 4. The Transmission of Work by
simple Machines—Chap. 5. Useful Propositions and Rules.

Breweries and Maltings : their Arrangement, Construction, Machinery, and Plant. By G. SCAMELL, F.R.I.B.A. Second
edition, revised, enlarged, and partly rewritten. By F. COLYER, M.I.C.E.,
M.I.M.E. *With* 20 *plates,* 8vo, cloth, 18s.

*A Practical Treatise on the Construction of Hori-
zontal and Vertical Waterwheels,* specially designed for the use of opera-
tive mechanics. By WILLIAM CULLEN, Millwright and Engineer. *With*
11 *plates.* Second edition, revised and enlarged, small 4to, cloth, 12s. 6d.

A Practical Treatise on Mill-gearing, Wheels, Shafts,
Riggers, etc. ; for the use of Engineers. By THOMAS BOX. Third
edition, *with* 11 *plates.* Crown 8vo, cloth, 7s. 6d.

Mining Machinery: a Descriptive Treatise on the
Machinery, Tools, and other Appliances used in Mining. By G. G.
ANDRÉ, F.G.S., Assoc. Inst. C.E., Mem. of the Society of Engineers.
Royal 4to, uniform with the Author's Treatise on Coal Mining, con-
taining 182 *plates,* accurately drawn to scale, with descriptive text, in
2 vols., cloth, 3l. 12s.

CONTENTS :

Machinery for Prospecting, Excavating, Hauling, and Hoisting—Ventilation—Pumping—
Treatment of Mineral Products, including Gold and Silver, Copper, Tin, and Lead, Iron
Coal, Sulphur, China Clay, Brick Earth, etc.

Tables for Setting out Curves for Railways, Canals,
Roads, etc., varying from a radius of five chains to three miles. By A.
KENNEDY and R. W. HACKWOOD. *Illustrated,* 32mo, cloth, 2s. 6d.

The Science and Art of the Manufacture of Portland

Cement, with observations on some of its constructive applications. With 66 *illustrations.* By HENRY REID, C.E., Author of 'A Practical Treatise on Concrete,' etc., etc. 8vo, cloth, 18*s.*

The Draughtsman's Handbook of Plan and Map

Drawing; including instructions for the preparation of Engineering, Architectural, and Mechanical Drawings. *With numerous illustrations in the text, and* 33 *plates* (15 *printed in colours*). By G. G. ANDRÉ, F.G.S., Assoc. Inst. C.E. 4to, cloth, 9*s.*

CONTENTS:

The Drawing Office and its Furnishings—Geometrical Problems—Lines, Dots, and their Combinations—Colours, Shading, Lettering, Bordering, and North Points—Scales—Plotting —Civil Engineers' and Surveyors' Plans—Map Drawing—Mechanical and Architectural Drawing—Copying and Reducing Trigonometrical Formulæ, etc., etc.

The Boiler-maker's and Iron Ship-builder's Companion,

comprising a series of original and carefully calculated tables, of the utmost utility to persons interested in the iron trades. By JAMES FODEN, author of 'Mechanical Tables,' etc. Second edition revised, *with illustrations,* crown 8vo, cloth, 5*s.*

Rock Blasting: a Practical Treatise on the means

employed in Blasting Rocks for Industrial Purposes. By G. G. ANDRÉ, F.G.S., Assoc. Inst. C.E. *With* 56 *illustrations and* 12 *plates,* 8vo, cloth, 10*s.* 6*d.*

Painting and Painters' Manual: a Book of Facts

for Painters and those who Use or Deal in Paint Materials. By C. L. CONDIT and J. SCHELLER. *Illustrated,* 8vo, cloth, 10*s.* 6*d.*

A Treatise on Ropemaking as practised in public and

private Rope-yards, with a Description of the Manufacture, Rules, Tables of Weights, etc., adapted to the Trade, Shipping, Mining, Railways, Builders, etc. By R. CHAPMAN, formerly foreman to Messrs. Huddart and Co., Limehouse, and late Master Ropemaker to H.M. Dockyard, Deptford. Second edition, 12mo, cloth, 3*s.*

Laxton's Builders' and Contractors' Tables; for the

use of Engineers, Architects, Surveyors, Builders, Land Agents, and others. Bricklayer, containing 22 tables, with nearly 30,000 calculations. 4to, cloth, 5*s.*

Laxton's Builders' and Contractors' Tables. Ex-

cavator, Earth, Land, Water, and Gas, containing 53 tables, with nearly 24,000 calculations. 4to, cloth, 5*s*

Sanitary Engineering: a Guide to the Construction
of Works of Sewerage and House Drainage, with Tables for facilitating
the calculations of the Engineer. By BALDWIN LATHAM, C.E., M. Inst.
C.E., F.G.S., F.M.S., Past-President of the Society of Engineers. Second
edition, *with numerous plates and woodcuts,* 8vo, cloth, 1*l.* 10*s.*

Screw Cutting Tables for Engineers and Machinists,
giving the values of the different trains of Wheels required to produce
Screws of any pitch, calculated by Lord Lindsay, M.P., F.R.S., F.R.A.S.,
etc. Cloth, oblong, 2*s.*

Screw Cutting Tables, for the use of Mechanical
Engineers, showing the proper arrangement of Wheels for cutting the
Threads of Screws of any required pitch, with a Table for making the
Universal Gas-pipe Threads and Taps. By W. A. MARTIN, Engineer.
Second edition, oblong, cloth, 1*s.*, or sewed, 6*d.*

A Treatise on a Practical Method of Designing Slide-
Valve Gears by Simple Geometrical Construction, based upon the principles
enunciated in Euclid's Elements, and comprising the various forms of
Plain Slide-Valve and Expansion Gearing ; together with Stephenson's,
Gooch's, and Allan's Link-Motions, as applied either to reversing or to
variable expansion combinations.   By EDWARD J. COWLING WELCH,
Memb. Inst. Mechanical Engineers. Crown 8vo, cloth, 6*s.*

Cleaning and Scouring: a Manual for Dyers, Laun-
dresses, and for Domestic Use. By S. CHRISTOPHER. 18mo, sewed, 6*d.*

A Glossary of Terms used in Coal Mining.   By
WILLIAM STUKELEY GRESLEY, Assoc. Mem. Inst. C.E., F.G.S., Member
of the North of England Institute of Mining Engineers. *Illustrated with*
numerous woodcuts and diagrams, crown 8vo, cloth, 5*s.*

A Pocket-Book for Boiler Makers and Steam Users,
comprising a variety of useful information for Employer and Workman,
Government Inspectors, Board of Trade Surveyors, Engineers in charge
of Works and Slips, Foremen of Manufactories, and the general Steam-
using Public. By MAURICE JOHN SEXTON. Second edition, royal
32mo, roan, gilt edges, 5*s.*

Electrolysis: a Practical Treatise on Nickeling,
Coppering, Gilding, Silvering, the Refining of Metals, and the treatment
of Ores by means of Electricity. By HIPPOLYTE FONTAINE, translated
from the French by J. A. BERLY, C.E., Assoc. S.T.E. *With engravings.*
8vo, cloth, 9*s.*

Barlow's Tables of Squares, Cubes, Square Roots,
Cube Roots, Reciprocals of all Integer Numbers up to 10,000. Post 8vo,
cloth, 6s.

A Practical Treatise on the Steam Engine, containing Plans and Arrangements of Details for Fixed Steam Engines,
with Essays on the Principles involved in Design and Construction. By
ARTHUR RIGG, Engineer, Member of the Society of Engineers and of
the Royal Institution of Great Britain. Demy 4to, *copiously illustrated
with woodcuts and* 96 *plates,* in one Volume, half-bound morocco, 2l. 2s.;
or cheaper edition, cloth, 25s.

This work is not, in any sense, an elementary treatise, or history of the steam engine, but
is intended to describe examples of Fixed Steam Engines without entering into the wide
domain of locomotive or marine practice. To this end illustrations will be given of the most
recent arrangements of Horizontal, Vertical, Beam, Pumping, Winding, Portable, Semi-
portable, Corliss, Allen, Compound, and other similar Engines, by the most eminent Firms in
Great Britain and America. The laws relating to the action and precautions to be observed
in the construction of the various details, such as Cylinders, Pistons, Piston-rods, Connecting-
rods, Cross-heads, Motion-blocks, Eccentrics, Simple, Expansion, Balanced, and Equilibrium
Slide-valves, and Valve-gearing will be minutely dealt with. In this connection will be found
articles upon the Velocity of Reciprocating Parts and the Mode of Applying the Indicator,
Heat and Expansion of Steam Governors, and the like. It is the writer's desire to draw
illustrations from every possible source, and give only those rules that present practice deems
correct.

A Practical Treatise on the Science of Land and
Engineering Surveying, Levelling, Estimating Quantities, etc., with a
general description of the several Instruments required for Surveying,
Levelling, Plotting, etc. By H. S. MERRETT. Fourth edition, revised
by G. W. USILL, Assoc. Mem. Inst. C.E. 41 *plates, with illustrations
and tables,* royal 8vo, cloth, 12s. 6d.

PRINCIPAL CONTENTS :

Part 1. Introduction and the Principles of Geometry. Part 2. Land Surveying; com-
prising General Observations—The Chain—Offsets Surveying by the Chain only—Surveying
Hilly Ground—To Survey an Estate or Parish by the Chain only—Surveying with the
Theodolite—Mining and Town Surveying—Railroad Surveying—Mapping—Division and
Laying out of Land—Observations on Enclosures—Plane Trigonometry. Part 3. Levelling—
Simple and Compound Levelling—The Level Book—Parliamentary Plan and Section—
Levelling with a Theodolite—Gradients—Wooden Curves—To Lay out a Railway Curve—
Setting out Widths. Part 4. Calculating Quantities generally for Estimates—Cuttings and
Embankments—Tunnels—Brickwork—Ironwork—Timber Measuring. Part 5. Description
and Use of Instruments in Surveying and Plotting—The Improved Dumpy Level—Troughton's
Level—The Prismatic Compass—Proportional Compass—Box Sextant—Vernier—Panta-
graph—Merrett's Improved Quadrant—Improved Computation Scale—The Diagonal Scale—
Straight Edge and Sector. Part 6. Logarithms of Numbers—Logarithmic Sines and
Co-Sines, Tangents and Co-Tangents—Natural Sines and Co-Sines—Tables for Earthwork,
for Setting out Curves, and for various Calculations, etc., etc., etc.

Health and Comfort in House Building, or Ventila-
tion with Warm Air by Self-Acting Suction Power, with Review of the
mode of Calculating the Draught in Hot-Air Flues, and with some actual
Experiments. By J. DRYSDALE, M.D., and J. W. HAYWARD, M.D.
Second edition, with Supplement, *with plates,* demy 8vo, cloth, 7s. 6d.

The Assayer's Manual: an Abridged Treatise on
the Docimastic Examination of Ores and Furnace and other Artificial
Products. By BRUNO KERL. Translated by W. T. BRANNT. *With* 65
illustrations, 8vo, cloth, 12*s*. 6*d*.

Electricity: its Theory, Sources, and Applications.
By J. T. SPRAGUE, M.S.T.E. Second edition, revised and enlarged, *with
numerous illustrations*, crown 8vo, cloth, 15*s*.

The Practice of Hand Turning in Wood, Ivory, Shell,
etc., with Instructions for Turning such Work in Metal as may be required
in the Practice of Turning in Wood, Ivory, etc. ; also an Appendix on
Ornamental Turning. (A book for beginners.) By FRANCIS CAMPIN.
Third edition, *with wood engravings*, crown 8vo, cloth, 6*s*.

CONTENTS :

On Lathes—Turning Tools—Turning Wood—Drilling—Screw Cutting—Miscellaneous
Apparatus and Processes—Turning Particular Forms—Staining—Polishing—Spinning Metals
—Materials—Ornamental Turning, etc.

Treatise on Watchwork, Past and Present. By the
Rev. H. L. NELTHROPP, M.A., F.S.A. *With* 32 *illustrations,* crown
8vo, cloth, 6*s*. 6*d*.

CONTENTS :

Definitions of Words and Terms used in Watchwork—Tools—Time—Historical Sum-
mary—On Calculations of the Numbers for Wheels and Pinions; their Proportional Sizes,
Trains, etc.—Of Dial Wheels, or Motion Work—Length of Time of Going without Winding
up—The Verge—The Horizontal—The Duplex—The Lever—The Chronometer—Repeating
Watches—Keyless Watches—The Pendulum, or Spiral Spring—Compensation—Jewelling of
Pivot Holes—Clerkenwell—Fallacies of the Trade—Incapacity of Workmen—How to Choose
and Use a Watch, etc.

Algebra Self-Taught. By W. P. HIGGS, M.A.,
D.Sc., LL.D., Assoc. Inst. C.E., Author of ' A Handbook of the Differential Calculus,' etc. Second edition, crown 8vo, cloth, 2*s*. 6*d*.

CONTENTS :

Symbols and the Signs of Operation—The Equation and the Unknown Quantity—
Positive and Negative Quantities—Multiplication—Involution—Exponents—Negative Expo-
nents—Roots, and the Use of Exponents as Logarithms—Logarithms—Tables of Logarithms
and Proportionate Parts—Transformation of System of Logarithms—Common Uses of
Common Logarithms—Compound Multiplication and the Binomial Theorem—Division,
Fractions, and Ratio—Continued Proportion—The Series and the Summation of the Series—
Limit of Series—Square and Cube Roots—Equations—List of Formulæ, etc.

Spons' Dictionary of Engineering, Civil, Mechanical,
Military, and Naval; with technical terms in French, German, Italian,
and Spanish, 3100 pp., and *nearly* 8000 *engravings*, in super-royal 8vo,
in 8 divisions, 5*l*. 8*s*. Complete in 3 vols., cloth, 5*l*. 5*s*. Bound in a
superior manner, half-morocco, top edge gilt, 3 vols., 6*l*. 12*s*.

Notes in Mechanical Engineering. Compiled principally for the use of the Students attending the Classes on this subject at the City of London College. By HENRY ADAMS, Mem. Inst. M.E., Mem. Inst. C.E., Mem. Soc. of Engineers. Crown 8vo, cloth, 2s. 6d.

Canoe and Boat Building: a complete Manual for Amateurs, containing plain and comprehensive directions for the construction of Canoes, Rowing and Sailing Boats, and Hunting Craft. By W. P. STEPHENS. *With numerous illustrations and 24 plates of Working Drawings.* Crown 8vo, cloth, 7s. 6d.

Proceedings of the National Conference of Electricians, *Philadelphia,* October 8th to 13th, 1884. 18mo, cloth, 3s.

Dynamo - Electricity, its Generation, Application, Transmission, Storage, and Measurement. By G. B. PRESCOTT. *With* 545 *illustrations.* 8vo, cloth, 1l. 1s.

Domestic Electricity for Amateurs. Translated from the French of E. HOSPITALIER, Editor of "L'Electricien," by C. J. WHARTON, Assoc. Soc. Tel. Eng. *Numerous illustrations.* Demy 8vo, cloth, 9s.

CONTENTS:

1. Production of the Electric Current—2. Electric Bells—3. Automatic Alarms—4. Domestic Telephones—5. Electric Clocks—6. Electric Lighters—7. Domestic Electric Lighting—8. Domestic Application of the Electric Light—9. Electric Motors—10. Electrical Locomotion—11. Electrotyping, Plating, and Gilding—12. Electric Recreations—13. Various applications—Workshop of the Electrician.

Wrinkles in Electric Lighting. By VINCENT STEPHEN. *With illustrations.* 18mo, cloth, 2s. 6d.

CONTENTS:

1. The Electric Current and its production by Chemical means—2. Production of Electric Currents by Mechanical means—3. Dynamo-Electric Machines—4. Electric Lamps—5. Lead—6. Ship Lighting.

The Practical Flax Spinner ; being a Description of the Growth, Manipulation, and Spinning of Flax and Tow. By LESLIE C. MARSHALL, of Belfast. *With illustrations.* 8vo, cloth, 15s.

Foundations and Foundation Walls for all classes of *Buildings,* Pile Driving, Building Stones and Bricks, Pier and Wall construction, Mortars, Limes, Cements, Concretes, Stuccos, &c. 64 *illustrations.* By G. T. POWELL and F. BAUMAN. 8vo, cloth, 10s. 6d.

Manual for Gas Engineering Students. By D. LEE.
18mo, cloth 1s.

Hydraulic Machinery, Past and Present. A Lecture
delivered to the London and Suburban Railway Officials' Association.
By H. ADAMS, Mem. Inst. C.E. *Folding plate.* 8vo, sewed, 1s.

Twenty Years with the Indicator. By THOMAS PRAY,
Jun., C.E., M.E., Member of the American Society of Civil Engineers.
2 vols., royal 8vo, cloth, 12s. 6d.

*Annual Statistical Report of the Secretary to the
Members of the Iron and Steel Association on the Home and Foreign Iron
and Steel Industries in 1884.* Issued March 1885. 8vo, sewed, 5s.

Bad Drains, and How to Test them; with Notes on
the Ventilation of Sewers, Drains, and Sanitary Fittings, and the Origin
and Transmission of Zymotic Disease. By R. HARRIS REEVES. Crown
8vo, cloth, 3s. 6d.

Well Sinking. The modern practice of Sinking
and Boring Wells, with geological considerations and examples of Wells.
By ERNEST SPON, Assoc. Mem. Inst. C.E., Mem. Soc. Eng., and of the
Franklin Inst., etc. Second edition, revised and enlarged. Crown 8vo,
cloth, 10s. 6d.

*Pneumatic Transmission of Messages and Parcels
between Paris and London, viâ Calais and Dover.* By J. B. BERLIER,
C.E. Small folio, sewed, 6d.

List of Tests (Reagents), arranged in alphabetical
order, according to the names of the originators. Designed especially
for the convenient reference of Chemists, Pharmacists, and Scientists.
By HANS M. WILDER. Crown 8vo, cloth, 4s. 6d.

*Ten Years' Experience in Works of Intermittent
Downward Filtration.* By J. BAILEY DENTON, Mem. Inst. C.E.
Second edition, with additions. Royal 8vo, sewed, 4s.

*A Treatise on the Manufacture of Soap and Candles,
Lubricants and Glycerin.* By W. LANT CARPENTER, B.A., B.Sc. (late
of Messrs. C. Thomas and Brothers, Bristol). *With illustrations.* Crown
8vo, cloth, 10s. 6d.

The Stability of Ships explained simply, and calculated
by a new Graphic method. By J. C. SPENCE, M.I.N.A. 4to, sewed,
3s. 6d.

Steam Making, or Boiler Practice. By CHARLES A.
SMITH, C.E. 8vo, cloth, 10s. 6d.

CONTENTS:

1. The Nature of Heat and the Properties of Steam—2. Combustion.—3. Externally Fired
Stationary Boilers—4. Internally Fired Stationary Boilers—5. Internally Fired Portable
Locomotive and Marine Boilers—6. Design, Construction, and Strength of Boilers—7. Pro-
portions of Heating Surface, Economic Evaporation, Explosions—8. Miscellaneous Boilers,
Choice of Boiler Fittings and Appurtenances.

The Fireman's Guide; a Handbook on the Care of
Boilers. By TEKNOLOG, föreningen T. I. Stockholm. Translated from
the third edition, and revised by KARL P. DAHLSTROM, M.E. Second
edition. Fcap. 8vo, cloth, 2s.

A Treatise on Modern Steam Engines and Boilers,
including Land Locomotive, and Marine Engines and Boilers, for the
use of Students. By FREDERICK COLYER, M. Inst. C.E., Mem. Inst. M.E.
With 36 *plates.* 4to, cloth, 25s.

CONTENTS:

1. Introduction—2. Original Engines—3. Boilers—4. High-Pressure Beam Engines—5.
Cornish Beam Engines—6. Horizontal Engines—7. Oscillating Fngines—8. Vertical High-
Pressure Engines—9. Special Engines—10. Portable Engines—11. Locomotive Engines—
12. Marine Engines

Steam Engine Management; a Treatise on the
Working and Management of Steam Boilers. By F. COLYER, M. Inst.
C.E., Mem. Inst. M.E. 18mo, cloth, 2s.

Land Surveying on the Meridian and Perpendicular
System. By WILLIAM PENMAN, C.E. 8vo, cloth, 8s. 6d.

The Topographer, his Instruments and Methods,
designed for the use of Students, Amateur Photographers, Surveyors,
Engineers, and all persons interested in the location and construction of
works based upon Topography. *Illustrated with numerous plates, maps,*
and engravings. By LEWIS M. HAUPT, A.M. 8vo, cloth, 18s.

A Text-Book of Tanning, embracing the Preparation
of all kinds of Leather. By HARRY R. PROCTOR, F.C.S., of Low Lights
Tanneries. *With illustrations.* Crown 8vo, cloth, 10s. 6d.

In super-royal 8vo, 1168 pp., *with* 2400 *illustrations*, in 3 Divisions, cloth, price 13*s*. 6*d*. each ; or 1 vol., cloth, 2*l*. ; or half-morocco, 2*l*. 8*s*.

A SUPPLEMENT

TO

SPONS' DICTIONARY OF ENGINEERING.

EDITED BY ERNEST SPON, MEMB. SOC. ENGINEERS.

Abacus, Counters, Speed Indicators, and Slide Rule.
Agricultural Implements and Machinery.
Air Compressors.
Animal Charcoal Machinery.
Antimony.
Axles and Axle-boxes.
Barn Machinery.
Belts and Belting.
Blasting. Boilers.
Brakes.
Brick Machinery.
Bridges.
Cages for Mines.
Calculus, Differential and Integral.
Canals.
Carpentry.
Cast Iron.
Cement, Concrete, Limes, and Mortar.
Chimney Shafts.
Coal Cleansing and Washing.

Coal Mining.
Coal Cutting Machines.
Coke Ovens. Copper.
Docks. Drainage.
Dredging Machinery.
Dynamo - Electric and Magneto-Electric Machines.
Dynamometers.
Electrical Engineering, Telegraphy, Electric Lighting and its practicaldetails,Telephones
Engines, Varieties of.
Explosives. Fans.
Founding, Moulding and the practical work of the Foundry.
Gas, Manufacture of.
Hammers, Steam and other Power.
Heat. Horse Power.
Hydraulics.
Hydro-geology.
Indicators. Iron.
Lifts, Hoists, and Elevators.

Lighthouses, Buoys, and Beacons.
Machine Tools.
Materials of Construction.
Meters.
Ores, Machinery and Processes employed to Dress.
Piers.
Pile Driving.
Pneumatic Transmission.
Pumps.
Pyrometers.
Road Locomotives.
Rock Drills.
Rolling Stock.
Sanitary Engineering.
Shafting.
Steel.
Steam Navvy.
Stone Machinery.
Tramways.
Well Sinking.

London: E. & F. N. SPON, 125, Strand.
New York: 12, Cortlandt Street.

Crown 8vo, cloth, with illustrations, 5s.

WORKSHOP RECEIPTS,

FIRST SERIES.

By ERNEST SPON.

SYNOPSIS OF CONTENTS.

Bookbinding.
Bronzes and Bronzing.
Candles.
Cement.
Cleaning.
Colourwashing.
Concretes.
Dipping Acids.
Drawing Office Details.
Drying Oils.
Dynamite.
Electro - Metallurgy — (Cleaning, Dipping, Scratch-brushing, Batteries, Baths, and Deposits of every description).
Enamels.
Engraving on Wood, Copper, Gold, Silver, Steel, and Stone.
Etching and Aqua Tint.
Firework Making — (Rockets, Stars, Rains, Gerbes, Jets, Tourbillons, Candles, Fires, Lances, Lights, Wheels, Fire-balloons, and minor Fireworks).
Fluxes.
Foundry Mixtures.

Freezing.
Fulminates.
Furniture Creams, Oils, Polishes, Lacquers, and Pastes.
Gilding.
Glass Cutting, Cleaning, Frosting, Drilling, Darkening, Bending, Staining, and Painting.
Glass Making.
Glues.
Gold.
Graining.
Gums.
Gun Cotton.
Gunpowder.
Horn Working.
Indiarubber.
Japans, Japanning, and kindred processes.
Lacquers.
Lathing.
Lubricants.
Marble Working.
Matches.
Mortars.
Nitro-Glycerine.
Oils.

Paper.
Paper Hanging.
Painting in Oils, in Water Colours, as well as Fresco, House, Transparency, Sign, and Carriage Painting.
Photography.
Plastering.
Polishes.
Pottery—(Clays, Bodies, Glazes, Colours, Oils, Stains, Fluxes, Enamels, and Lustres).
Scouring.
Silvering.
Soap.
Solders.
Tanning.
Taxidermy.
Tempering Metals.
Treating Horn, Mother-o'-Pearl, and like substances.
Varnishes, Manufacture and Use of.
Veneering.
Washing.
Waterproofing.
Welding.

Besides Receipts relating to the lesser Technological matters and processes, such as the manufacture and use of Stencil Plates, Blacking, Crayons, Paste, Putty, Wax, Size, Alloys, Catgut, Tunbridge Ware, Picture Frame and Architectural Mouldings, Compos, Cameos, and others too numerous to mention.

London: E. & F. N. SPON, 125, Strand.

New York: 12, Cortlandt Street.

Crown 8vo, cloth, 485 pages, with illustrations, 5*s.*

WORKSHOP RECEIPTS,

SECOND SERIES.

By ROBERT HALDANE.

Pigments, Paint, and Painting : embracing the preparation of *Pigments*, including alumina lakes, blacks (animal, bone, Frankfort, ivory, lamp, sight, soot), blues (antimony, Antwerp, cobalt, cæruleum, Egyptian, manganate, Paris, Péligot, Prussian, smalt, ultramarine), browns (bistre, hinau, sepia, sienna, umber, Vandyke), greens (baryta, Brighton, Brunswick, chrome, cobalt, Douglas, emerald, manganese, mitis, mountain, Prussian, sap, Scheele's, Schweinfurth, titanium, verdigris, zinc), reds (Brazilwood lake, carminated lake, carmine, Cassius purple, cobalt pink, cochineal lake, colcothar, Indian red, madder lake, red chalk, red lead, vermilion), whites (alum, baryta, Chinese, lead sulphate, white lead—by American, Dutch, French, German, Kremnitz, and Pattinson processes, precautions in making, and composition of commercial samples—whiting, Wilkinson's white, zinc white), yellows (chrome, gamboge, Naples, orpiment, realgar, yellow lakes) ; *Paint* (vehicles, testing oils, driers, grinding, storing, applying, priming, drying, filling, coats, brushes, surface, water-colours, removing smell, discoloration ; miscellaneous paints—cement paint for carton-pierre, copper paint, gold paint, iron paint, lime paints, silicated paints, steatite paint, transparent paints, tungsten paints, window paint, zinc paints) ; *Painting* (general instructions, proportions of ingredients, measuring paint work ; carriage painting—priming paint, best putty, finishing colour, cause of cracking, mixing the paints, oils, driers, and colours, varnishing, importance of washing vehicles, re-varnishing, how to dry paint ; woodwork painting).

London: E. & F. N. SPON, 125, Strand.
New York: 12, Cortlandt Street.

WORKSHOP RECEIPTS,
FOURTH SERIES,
DEVOTED MAINLY TO HANDICRAFTS & MECHANICAL SUBJECTS.
By C. G. WARNFORD LOCK.

250 Illustrations, with Complete Index, and a General Index to the Four Series, 5s.

Waterproofing — rubber goods, cuprammonium processes, miscellaneous preparations.

Packing and Storing articles of delicate odour or colour, of a deliquescent character, liable to ignition, apt to suffer from insects or damp, or easily broken.

Embalming and Preserving anatomical specimens.

Leather Polishes.

Cooling Air and Water, producing low temperatures, making ice, cooling syrups and solutions, and separating salts from liquors by refrigeration.

Pumps and Siphons, embracing every useful contrivance for raising and supplying water on a moderate scale, and moving corrosive, tenacious, and other liquids.

Desiccating—air- and water-ovens, and other appliances for drying natural and artificial products.

Distilling—water, tinctures, extracts, pharmaceutical preparations, essences, perfumes, and alcoholic liquids.

Emulsifying as required by pharmacists and photographers.

Evaporating—saline and other solutions, and liquids demanding special precautions.

Filtering—water, and solutions of various kinds.

Percolating and Macerating.

Electrotyping.

Stereotyping by both plaster and paper processes.

Bookbinding in all its details.

Straw Plaiting and the fabrication of baskets, matting, etc.

Musical Instruments—the preservation, tuning, and repair of pianos, harmoniums, musical boxes, etc.

Clock and Watch Mending—adapted for intelligent amateurs.

Photography—recent development in rapid processes, handy apparatus, numerous recipes for sensitizing and developing solutions, and applications to modern illustrative purposes.

London: E. & F. N. SPON, 125, Strand.
New York: 12, Cortlandt Street.

In demy 8vo, cloth, 600 pages, and 1420 Illustrations, 6s.

SPONS'
MECHANICS' OWN BOOK;
A MANUAL FOR HANDICRAFTSMEN AND AMATEURS.

CONTENTS.

Mechanical Drawing—Casting and Founding in Iron, Brass, Bronze, and other Alloys—Forging and Finishing Iron—Sheetmetal Working—Soldering, Brazing, and Burning—Carpentry and Joinery, embracing descriptions of some 400 Woods, over 200 Illustrations of Tools and their uses, Explanations (with Diagrams) of 116 joints and hinges, and Details of Construction of Workshop appliances, rough furniture, Garden and Yard Erections, and House Building—Cabinet-Making and Veneering — Carving and Fretcutting — Upholstery — Painting, Graining, and Marbling — Staining Furniture, Woods, Floors, and Fittings—Gilding, dead and bright, on various grounds—Polishing Marble, Metals, and Wood—Varnishing—Mechanical movements, illustrating contrivances for transmitting motion—Turning in Wood and Metals—Masonry, embracing Stonework, Brickwork, Terracotta, and Concrete—Roofing with Thatch, Tiles, Slates, Felt, Zinc, &c.—Glazing with and without putty, and lead glazing—Plastering and Whitewashing—Paper-hanging—Gas-fitting—Bell-hanging, ordinary and electric Systems — Lighting — Warming — Ventilating — Roads, Pavements, and Bridges — Hedges, Ditches, and Drains — Water Supply and Sanitation—Hints on House Construction suited to new countries.

London: E. & F. N. SPON, 125, Strand.
New York: 12, Cortlandt Street.

www.ingramcontent.com/pod-product-compliance
Lightning Source LLC
Chambersburg PA
CBHW021939190326
41519CB00009B/1073